Charles Seale-Hayne Library
University of Plymouth
(01752) 588 588
LibraryandITenquiries@plymouth.ac.uk

FORMATION AND DEFORMATION OF GLACIAL DEPOSITS

PROCEEDINGS OF THE MEETING OF THE COMMISSION ON THE FORMATION
AND DEFORMATION OF GLACIAL DEPOSITS / DUBLIN / IRELAND / MAY 1991

Formation and Deformation of Glacial Deposits

Edited by
WILLIAM P.WARREN
Geological Survey of Ireland, Dublin
DAVID G.CROOT
University of Plymouth, UK

A.A.BALKEMA / ROTTERDAM / BROOKFIELD / 1994

The texts of the various papers in this volume were set individually by typists under the supervision of each of the authors concerned.

Authorization to photocopy items for internal or personal use, or the internal or personal use of specific clients, is granted by A.A. Balkema, Rotterdam, provided that the base fee of US$1.00 per copy, plus US$0.10 per page is paid directly to Copyright Clearance Center, 222 Rosewood Drive, Danvers, MA 01923, USA. For those organizations that have been granted a photocopy license by CCC, a separate system of payment has been arranged. The fee code for users of the Transactional Reporting Service is: 90 5410 096 6/94 US$1.00 + US$0.10.

Published by
A.A. Balkema, P.O. Box 1675, 3000 BR Rotterdam, Netherlands
A.A. Balkema Publishers, Old Post Road, Brookfield, VT 05036, USA

ISBN 90 5410 096 6
© 1994 A.A. Balkema, Rotterdam
Printed in the Netherlands

Formation and Deformation of Glacial Deposits, Warren & Croot (eds) © 1994 Balkema, Rotterdam, ISBN 90 5410 096 6

Table of contents

Interpreting the glacial record

Formation and Deformation of Glacial Deposits, Warren & Croot (eds) © 1994 Balkema, Rotterdam, ISBN 90 5410 096 6

Introduction

The INQUA Commission on the Formation and Properties of Glacial Deposits met in Ireland in May 1991 with a fourfold objective: to examine approaches to mapping glacigenic sediments; to discuss glacial depositional processes; to consider glaciotectonic structures and processes; and to do this in the context of Irish glacial sediments, the complexity of which has resulted in radically different interpretations during the last 150 years. The meeting was organised jointly by the Work Group on Mapping Glacial Deposits and the Work Group on Glacial Tectonics.

The first part of the meeting was concerned with examining sediments on the east and south coasts in the context of conflicting interpretations as to genesis, the chief question being 'Are they primarily glaciomarine or terrestrial glacial deposits?' Coastal exposures at Killiney, south of Dublin, Blackwater, County Wexford and Kilmore Quay on the south coast of Wexford exhibited strong evidence of sub-glacial tectonic processes. Questions still remain as to whether some of the sediments (both diamictons and laminated sequences) were originally glaciomarine, and as to the extent they might subsequently have been resedimented or reconstituted by sub-glacial processes.

An opportunity was taken to view undisputed terrestrial glacigenic sediments in the foreland of the mountains of the southwest where the glacial geomorphology is clear.

This provided the backdrop to a two-day seminar held in Dingle, Co. Kerry. This volume contains fifteen of the papers which formed the chief focus for discussion. Abstracts of all papers presented were published at the time of the symposium. Some of the full papers not published here had already been published whilst others have since been published elsewhere.

The contents of this volume reflect the main themes of the symposium. A number of presentations concerned with approaches to mapping glacial sediments are not included. These might more appropriately form the basis for a technical volume dedicated to this topic.

The papers in this volume encapsulate the work of the sponsoring INQUA Commission and its predecessor, focusing scientific analysis on questions of till genesis and the interpretation of depositional processes and environments. In this respect the papers are the fruits of both the formal and informal activities of the past and present Work Groups. We are confident that the stimulating presentations and lively discussion witnessed at Dingle will encourage further research and yet more challenging hypotheses.

Following the symposium, the group travelled north to examine, *inter alia*, sedimentary and glaciotectonic structures of a drumlin in Galway Bay and later to see spectacular glaciotectonic folds in Carboniferous shales and limestones on the south coast of Donegal Bay. The criteria for distinguishing glaciomarine sediments were again addressed when examining shell-bearing diamictons and gravels further west along the same coast.

The return journey to Dublin across the Central Plain facilitated discussion on the genesis of the large esker systems that traverse the plain. Extensive glaciolacustrine sediments were used to demonstrate a new interpretation of the Irish ice sheet as one which developed from a number of coalescing ice domes. These then acted as centres towards which the ice margins retreated during deglaciation. This led to ice damming of the whole Central Plain area and deposition of complex linear ice marginal subaqueous fans in association with tunnel esker deposits. Thus many of the features which gave the name esker to the geological literature are seen to be more appropriately termed moraines. However given that many such features are cored by tunnel esker deposits, we had once again seen evidence for the immense complexity of glacigenic sedimentary processes. These features which had in recent times been variously interpreted as glaciomarine, subaerial moraines, ice tunnel deposits and glacial crevasse fillings have found an interpretation which accommodates the principles upon which all of these interpretations were based. Can we now find processes which will reconcile the differences between those who see glaciomarine sediments where others see direct terrestrially based glacial deposits?

We are confident the authors would agree that many of the papers included in this volume are in large measure products of similar discussions in the field. These discussions represent the most important activities of the Work Groups. The scientific enrichment experienced in these international discussions finds it linguistic counterpart in the idiomatic variation that must pervade a book such as this in which the majority of papers are authored by people whose first language is not English. This characteristic should not be confused with grammatical or typographical errors, any of which remain the responsibility of the editors.

William P. Warren
David G. Croot

Formation of diamictons (glacial versus non-glacial)

Formation and Deformation of Glacial Deposits, Warren & Croot (eds) © 1994 Balkema, Rotterdam, ISBN 90 5410 096 6

Waterlain and lodgement till facies of the lower sedimentary complex from the Dänischer-Wohld Cliff, Schleswig-Holstein, North Germany

Jan A. Piotrowski
Institute of Geology and Palaeontology, University of Kiel, Germany

ABSTRACT: Examples of the sediment facies from the lower sedimentary complex along the Dänischer-Wohld-Cliff are presented. The first major facies comprises waterlain tills deposited primarily in a proglacial lake formed between the advancing Weichselian ice sheet and the Saalian highlands. The second facies is a lodgement till, deposited as the glacier overrode the waterlain sequence. The whole complex is attributed to a single ice advance.

1 INTRODUCTION

Good exposures along the Baltic Sea cliff between Eckernförde and Kiel provide insight into glacial deposits over a distance of some 30 km (Fig.1). This part of Schleswig-Holstein, which extends into the Baltic Sea between deep inland incisions of the Kieler Förde and the Eckernförde Bight, contain a remarkable inventory of different glacigenic facies and depositional milieus of the Weichselian Glaciation. The exposed sequence was subdivided by Piotrowski (1989 and in press) into two complexes. The upper complex, consisting of a succession of lodgement, melt-out and flow tills derives from the youngest Weichselian ice advance (the Fehmarn advance) and was described in detail by Piotrowski (in press). The lower complex corresponds to the first four Weichselian advances distinguished by Stephan & Menke (1977) and Stephan et al. (1983). The purpose of this paper is to give examples of two major diamicton facies of the lower complex and to suggest a palaeogeomorphological development of the area at that time. A more detailed account on this topic involving fabric measurements, petrographic analysis and glaciotectonic measurements is in preparation.

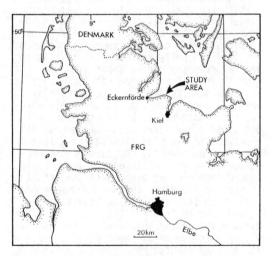

Fig.1 Location of the examined cliff sections

2 THE FIRST FACIES

The first facies is the most wide-spread diamicton occupying the lower cliff portion from about sea level to some 3-4 m upwards. It can be traced almost continously along the well-exposed cliff sections and, because of its distinct structure, can be positively identified as one sedimentary unit. The sediment is a relatively coarse-textured diamicton with low clay content and abundance of gravel-sized stones. Its characteristic feature is light-coloured mm to dm thick stringers and flat lenses of chalk-enriched material highlighted against the otherwise olive-grey matrix (Fig.2). In places where the chalk content is very high the diamicton appears to be white-painted to form alternating light and dark stripes. Individual laminae are often several metres long, and merge with one another or wedge out

Fig.2 Chalk-rich, laminated diamicton
(waterlain flow till)

Fig.3 Bullet-shaped sand clast resting on a
chalk-rich laminae embedded in diamicton.
Note the deformation in front of the clast
caused by its falling onto a plastic, saturated substratum of a water-filled basin.
The diamicton is a waterlain flow till.

into the grey matrix. Resistant stones
usually lack signs of extensive abrasion,
and are typically angular and unstriated.
The chalk clusters are often so fragile
that they break easily between fingers.

Where the chalk-rich layers collide with
larger stones, small folds occur which are
downwarped below the stones and form convex
coatings on top of them. These disturbances
are longitudinally asymmetrical indicating
lateral dragging during deposition or
shortly thereafter. Symmetrical coatings or
let-down structures caused by primarily
vertical deposition and sediment-volume reduction considered to be typical for melt-
out tills (Shaw 1979) were only occasionally
observed. An example illustrating a lateral
movement component at the time of sedimentation is shown in Figure 3. A large bullet-
shaped sand cluster blunted to the left and
tapered on the right rests on a downwarped
chalk-rich layer. The layer itself is
squeezed up in front of the blunt end of the
cluster indicating that the cluster, probably in a frozen or at least partly frozen
state re-sedimented onto the chalk layer on
top of the palaeo-surface sloping to the
left. It was subsequently buried under
younger portions of the diamicton. The shape
of the cluster results from streamlining
during the re-sedimentation prior to the
final deposition on the chalk layer.

Closely associated with these directional
structures are minute, flat, isoclinal folds
rarely exceeding 20-30 cm in amplitude.
They consist in many cases of thin sorted
sediments layers or chalky chorizons. The
folds are never underlain or capped by
unequivocally erosional disconformities.
Instead, the amplitude of individual folds
gradually diminishes away from the fold
axes   into nondeformed matrix. They often
occur as convolute bedding systems resting
on inclined palaeo-surfaces.

Furthermore, the diamicton facies is often
separated by layers of waterlain, sorted
sediments with a dominance of silt, clay and
fine-grained sand. Thickness of the layers varies from few mm to about 2 m and in places
volumetrically exceeds the diamicton and
thus constitute the bulk of the sequence.
Silts and fine sands are distinctly bedded,
whereby minute climbing ripples and horizontal laminations dominate (Fig.4).
Throughout the sequences of alternating
sorted sediments and diamicton sheets a
striking feature is again a complete absence
of erosional contacts. The sediments either
grade into one another or where the grain-
size contacts are sharp, the base of the
overlying unit rests conformably on the
micro-relief of the underlying unit. An
example is given in Figure 5 depicting an
about 15 cm thick diamicton layer sandwiched
between silty-sand beds. The base of the
diamicton traces exactly the rippled surface
below, which could be observed for some tens
of cm into the cliff face after carefully
removing the sand. This reveals that diamicton sedimentation was a subaqueous, low-
energy, particle-by-particle rain-out process. H.-J.Stephan (personal communication)
considers also the possibility of flow till
sedimentation onto a frozen lake surface,
from which the till would be subsequently
lowered down onto substratum upon ice melting. Where the diamicton is intercalated
with fine-laminated clays (occasionally
occurring as varved clays), the contacts
are also gradational or conformable with no
sign of erosional activity (Fig.6).

The sedimentary coexistence of sorted
layers and the diamicton indicates that
both were deposited in a common, aqueous

Fig.4 Diamicton (waterlain melt-out till) conformably interbedded with rippled fine sand and silt

Fig.6 Clay (partly varvad), silt and fine sand resting within a diamicton (waterlain till). Note depositional contacts throughout the sediment succession

Fig.5 A section of Figure 4 showing diamicton (waterlain melt-out till) sandwiched between fine sand and silt beds. Note the distinctly conformable contact at the base of the diamicton

sedimentary milieu. A cold environment is documented by numerous apparently ice-rafted, oversized cobbles and boulders associated with sorted layers. The elongated stones are typically oriented at high angles to bedding planes and downwarping of substratum caused by a drop-impact is common. A large drop-stone shown in Figure 7 is covered by undisturbed sediment layers which indicates that it must have sunk into the substratum relatively rapidly, before being covered by younger deposits. Elongated, unstriated boulders resting subvertically in the diamicton matrix are probably drop-stones as well.

In numerous sections the whole sequence is disturbed into large-scale folds of amplitude often exceeding 10 m (cf. Seifert 1954 and Prange 1987). In contrast to small-

scale convolute bedding and similar synsedimentary disturbances mentioned above, the large folds postdate the formation of the sequence (see below).

For interpretation of depositional processes involved in the origin of the diamicton the following evidence is considered crucial:

1. synsedimentary flow-structures,
2. bullet-shaped clusters of unconsolidated materials with stoss-side deformations in the substratum,
3. ice-rafted stones,
4. abundance of angular boulders,
5. frequent and conformable intercalations of waterlaid materials,
6. pronounced lack of erosional disconformities and
7. presence of erratic clasts.

These features indicate that deposition took place in a water-filled basin in proximity to a glacier that served as sediment source. Depending on the ice thawing rate, the supply of morainic material and its grain-size, and the hydraulic characteristics of the basin, either diamicton or sorted sediments were laid down. Where small-scale deformation structures reveal lateral movement component, the deposition very likely proceeded on the dipping slope of the basin (marginal, probably proximal basin sections). Where no such features occur, deposition involved rain-out of diamict matrix and ice-rafted stones onto a horizontal or only slightly dipping surface (central basin sections). Both proglacial sedimentation in front of a calving ice margin and subglacial sedimentation as a result of undermelt release of morainic material are plausible. Because of the large lateral extent of this facies the former

Fig. 7 A large boulder (ice-rafted drop-stone) resting on downwarped sand bed and covered by undisturbed succession of sand, silt, clay and diamicton layers.

Fig. 8 Contact zone between disturbed sands and diamicton. The sands are heavily deformed and truncated right below the diamicton. The diamicton is a lodgement till whose basal 5 cm consist of pronouncely sheared section with extremly thin stringers of sand derived from the substratum. This basal section is called here "syndepositionally sheared till".

probably dominated. Bearing in mind that during the Weichselian Glaciation ice advanced in the study area out of the Baltic Sea depression against the Saalian highlands, a narrow glacial lake running parallel to the ice margin can be easily conceived.

Putting aside the ongoing discussion on terminology of glacigenic sediments (e.g. Dreimanis 1988), the diamicton in question can be genetically described as a combination of waterlain flow till (cf. Evenson et al. 1977) and waterlain melt-out till (cf. Gibbard 1980).

Worthwhile to note is a striking structural resemblance of the chalk-enriched laminations within this facies to the so-called "chalk till" from the Brodtener Cliff some 70 km to the southeast (Gagel 1915, Kabel 1987), to stratified till on the Isle of Lyö in Denmark interpreted as flow till (Marcussen 1973) and to laminated pebbly mud of the North Sea Drift interpreted by Eyles et al. (1989) as a product of sub-aqueous deposition in front of the margin of an ice sheet terminating in a large water body. Sections of the diamicton from the Brodtener Cliff are interpreted by Ehlers (1990 p.25) as a subglacial waterlain till.

3 THE SECOND FACIES

The second distinct facies of the lower complex exposed on the Dänischer-Wohld-Cliff is a primarily massive, several metres thick fine-grained diamicton with a relatively low boulder content. Pebbles are often striated and in a few places a sub-horizontal imbrication was noticed. The diamicton either directly overlies the waterlain till and in such places a precise boundary between the two is difficult to delineate, or it caps sands occassionally found on top of the waterlain till. Where the latter is the case, the sands are always truncated by an erosional unconformity at the base of the diamicton, especially apparent where the original bedding in sand is preserved. Figure 8 shows a contact zone between a highly contorted sand and the diamicton. The intensity of disturbances within sand increases upwards and the uppermost 5 cm immediately below the diamicton is intensely sheared and displaced to the right (west). The basal zone of the diamicton is intensely sheared as well and incorporates very thin stringers of sand from below.

The diamicton is interpreted as a lodgement till sensu Dreimanis (1988), the deposition of which was accompanied by glaciotectonic disturbances in the underlying sediments. Disturbances were facilitated by a high pore water content as documented by the plastic, non-brittle character of deformations. Considering a high intensity of shearing within the basal zone of the till, associated with incorporation of the substratum material and its reworking into extremly flat, elongated and base-parallel stringers, the term "syndepositionally sheared till" can be applied to this till zone. The adjective "syndepositionally" is used here to distinguish the basal shear planes from other possible shear planes in tills which can originate, for instance, due to postdepositional overriding by a younger ice advance. Similar sequence of glaciofluvial sands and fine-grained rhythmites deformed in the shear zone below till was described by van der Meer et al. (1985, their unit C-1) and considered analogous to Banham's (1977) zone C of penetratively sheared sediments. Stephan (1988) called diamictons originating by pure shear a "shear till" and gave examples from several localities in North Germany which he interpreted as having been formed beneath cold glaciers. It is believed that deformations of the kind shown in Figure 8, typical for the Dänischer-Wohld-Cliff, originated under a temperate glacier overriding water-saturated sediments (cf. Boulton & Jones 1979).

The lodgement till is in places heavily folded together with the waterlain till, which points out that the major folding phase took place after the whole lower sedimentary complex had been deposited. In several localities, however, the lodgement till was found to truncate small folds of older deposits, which documents a minor deformation phase predating the emplacement of this till.

4 PALAEOENVIRONMENTAL RECONSTRUCTION AND CONCLUSIONS

Based on the data presented the events leading to the formation of the lower sedimentary complex can be summarised as follows:

1. In front of the advancing Weichselian ice sheet a proglacial lake, or a system of lakes, developed due to trapping of meltwaters and possibly also fluvial waters in depression between the ice margin and palaeorelief sloping icewards. This triggered sedimentation of waterlain flow tills and melt-out tills. Boulders were deposited primarily as drop-stones melting from ice rafts, and the matrix either as subaqueous mudflows slumping into the basin at the ice margin or as a result of gradual particle rain-out from suspension some distance from the glacier. Till deposition was periodically interrupted by sedimentation of fine-grained sorted material under low-energy hydraulic conditions. Minor syndepositional folding triggered by sediment re-mobilisation and downwards creeping occurred. It is possible that the deposition also took place partly in subglacial water-filled basins, although the apparent continuity of this facies for many kilometers suggests predominantly proglacial aqueous environment. Considering the up-ice sloping palaeorelief and the depth of the Baltic depression, the ice margin was probably grounded most of the time.

2. Simultaneous with the progressing ice advance the proglacial lake was "pushed" further to the south while finally the glacier overrode the waterlain sequence exposed today along the cliff sections. This was accompanied by minor glaciotectonic disturbances, truncation of the substratum and intensive shearing at the ice base, all favoured by high water pressure of saturated sediments. Subsequently a massive lodgement till with intensely sheared basal part was emplaced as the glacier advance proceeded. Ice margin fluctuations described by Stephan & Menke (1977) as advances 2 to 4 took place further to the south so that the study area remained ice-covered until the end of the 4th advance. This is indicated by a lack of erosional disconformities which could be correlated to ice retreat phases.

3. As the study area was located again in the ice-marginal zone during the ice retreat, major glaciotectonic deformations occurred leading to the formation of high-amplitude folds. This deformation phase is documented by the fact that the waterlain till and the lodgement till likewise are found as an en masse folded sediment sequence. Before the last ice-overriding during the Late-glacial Fehmarn Advance (see Prange 1987) during which lodgement tills and an extensive cover of ablation till were emplaced (Piotrowski, in press), the area under discussion remained ice-free.

ACKNOWLEDGEMENTS

Gratitude is extended to Dr H.-J. Stephan for valuable discussions and to Dr J. van der Meer for critically reviewing the manuscript.

REFERENCES

Banham, P.H. 1977. Glaciotectonics in till stratigraphy. Boreas 6:101-105.
Boulton, G.S. & A.S.Jones 1979. Stability of temperate ice caps and ice sheets resting on beds of deformable sediment. J. Glaciol. 24:29-43.
Dreimanis, A. 1988. Tills: their genetic terminology and classification. In R.P. Goldthwait & C.L.Matsch (eds.), Genetic classification of glacigenic deposits, p.17-83. Rotterdam, Balkema.
Ehlers, J. 1990. Untersuchungen zur Morphodynamik der Vereisungen Norddeutschlands unter Berücksichtigung benachbarter Gebiete. Bremer Beiträge zur Geographie und Raumplanung 19:1-166.
Evenson, E.B., A.Dreimanis & J.W.Newsome 1977. Subaquatic flow tills: a new interpretation for the genesis of some laminated till deposits. Boreas 6:115-133.
Eyles, N., C.H.Eyles & A.M.McCabe 1989. Sedimentation in an ice-contact subaquous setting: the mid-Pleistocene "North Sea

Drifts" of Norfolk. U.K. Quat. Sci. Rev. 8:57-74.

Gagel, C. 1915. Erläuterungen zur geologischen Karte von Preussen und angrenzender Bundesstaaten, Blatt Curau-Schwartau-Travemünde. Kgl. preuss. geol. Landesanstalt.

Gibbard, P. 1980. The origin of stratified Catfish Creek Till by basal melting. Boreas 9:71-85.

Kabel, K. 1987. Petrographical and structural investigations of the Brodtener Ufer cliff. In J.J.M. van der Meer (ed.), Tills and glaciotectonics, p.89-96. Rotterdam, Balkema.

Marcussen, I. 1973. Studies on flow till in Denmark. Boreas 2:213-231.

van der Meer, J.J.M., M.Rappol & J.Semeijn 1985. Sedimentology and genesis of glacial deposits in the Goudsberg, central Netherlands. Mededelingen Rijks geol. Dienst 39-2:2-29.

Piotrowski, J.A. 1989. A melt-out sequence from Dänischer Wohld, Schleswig-Holstein, as palaeogeomorphological indicator. II. Int. Conf. on Geomorphology, Frankfurt am Main, Abstracts: 223-224.

Piotrowski, J.A. (in press). Till facies and depositional environments of the upper sedimentary complex from the Stohler Cliff, Schleswig-Holstein, North Germany. Zeitschrift für Geomorphologie.

Prange, W. 1987. Gefügekundliche Untersuchungen der weichselzeitlichen Ablagerungen an den Steilufern des Dänischen Wohlds, Schleswig-Holstein. Meyniana 39: 85-110.

Seifert, G. 1954. Das mikroskopische Korngefüge des Geschiebemergels als Abbild der Eisbewegung, zugleich Geschichte des Eisabbaues in Fehmarn, Ost-Wagrien und dem Dänischen Wohld. Meyniana 2:124-190.

Shaw, J. 1979. Genesis of the Sveg tills and Rogen moraines of central Sweden: a model of basal melt out. Boreas 8: 409-426.

Stephan, H.-J. 1988. Origin of a till-like diamicton by shearing. In R.P.Goldthwait & C.L.Matsch (eds.), Genetic classification of glacigenic deposits, p.93-96. Rotterdam, Balkema.

Stephan, H.-J. & B.Menke 1977. Untersuchungen über den Verlauf der Weichsel-Kaltzeit in Schleswig-Holstein. Zeitschrift für Geomorphologie, Suppl. Bd. 27:12-28.

Stephan, H.-J., C.Kabel & G.Schlüter 1983. Stratigraphical problems in the glacial deposits of Schleswig-Holstein. In J.Ehlers (ed.), Glacial deposits in north-west Europe, p.305-320. Rotterdam, Balkema.

Formation and Deformation of Glacial Deposits, Warren & Croot (eds) © 1994 Balkema, Rotterdam, ISBN 90 5410 096 6

Glacial and non-glacial diamictons in the Karakoram Mountains and Western Himalayas

Lewis A.Owen
Royal Holloway and Bedford New College, UK

ABSTRACT: There are vast thicknesses of glacial and non-glacial diamictons in the Karakoram Mountains and western Himalayas. These are deposited by a combination of gravity controlled processes (mass movement) and by a highly active glacial system. Supraglacial tills and tills resedimented by mass movement processes dominate the landscape and the sedimentary record. These show a large variety of sedimentary and geomorphological features which are described in detail. This information can be used to help elucidate the genesis of problematic deposits in this and other similar high mountain regions. This is critical for reconstructing accurate palaeoenvironments. Methods of study include: field mapping and logging; particle size analysis; macro-and microfabrics analysis; clay mineralogy; geotechnical analysis; and micromorphology. Emphasis is placed on micromorphological studies which provide important new information on the sedimentology of diamictons.

1 INTRODUCTION

The Karakoram and western Himalayan mountains are one of the most geodynamically active regions on earth, with some of the longest glaciers outside the polar region and some of the world's largest rivers including the Indus, Gilgit and Hunza (Figure 1). The region contains some of the highest mountains with peaks rising between 7000 and 8000m and includes K2 (8611m). This landscape is the result of a combination of rapid uplift produced by the inter-continental collision of the Indian and Asian plates (Gansser, 1964; Dewey and Burke, 1973; Le Fort, 1975), and intense denudation, involving frost shattering (Hewitt, 1968; Goudie et al. 1984); chemical weathering by salt crystal growth (Goudie, 1984; Whalley et al. 1984) and granular disintegration (Goudie et al. 1984); glacial erosion (Goudie et al. 1984; Li Jijun et al. 1984); fluvial incision (Ferguson, 1984; Ferguson et al. 1984) and mass movement (Brunsden and Jones, 1984; Brunsden et al. 1983). Diamictons of glacial, glaciofluvial, and mass movement origin are widespread in the valleys of the Karakoram Mountains and western Himalayas, forming an important component of the valley fill sediments. These valley fills are impressive, often exceeding 500m in thickness and are commonly over 100m thick. These represent contemporary glacial and paraglacial deposits, remnants of past glaciations and recent glacial advances (Derbyshire et al. 1984; Owen, 1988a).

There have been few sedimentological studies in the Himalayan mountain ranges, and rarely any concerning the sedimentology of glacial and non-glacial diamictons. Studies have concentrated on western end of this mountain belt, in northern Pakistan (Derbyshire, 1984; Li et al. 1984; Owen, 1988a, 1988b and 1989; Owen and Derbyshire, 1988 and 1989; Derbyshire and Owen, 1990), where access was made easier with the completion in 1980 of the Karakoram Highway. This traverses the grain of the mountains connecting Pakistan and China, and reaches altitudes of up to to 5528m. Diamictons in this region show considerable variability, but variations in particle size and shape, pebble anisotropy, microfabrics, thickness, morphology, weathering characteristics, geotechnical properties and facies associations are of particular importance in diagnosing the genesis of these deposits. The intense fluvial and glacial erosion has dissected and destroyed diagnostic morphologies in many places and tills are frequently confused with debris flow deposits and vice versa. This is particularly problematical when mass movement material contains facetted and straited clast i.e. derived either directly or indirectly from glacial ice or from reworked and resedimented tills. Till derived from sediment which has been transported only within passive flow paths within very steep glaciers may have unstraited clasts, confusing the matter further.

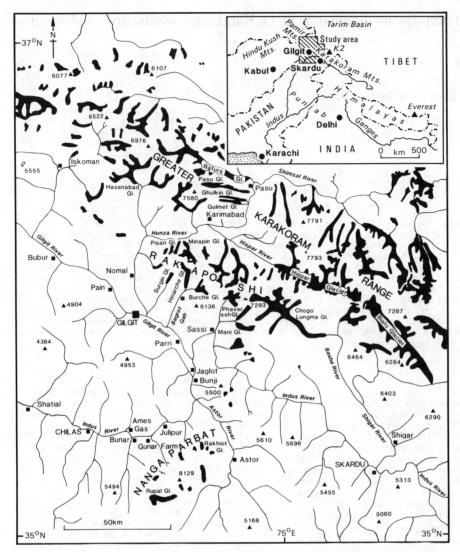

Figure 1. Location map of the Karakoram Mountains and western Himalayas

Glacially derived facies constitute a gradational series based on their particle size characteristics, but the debris flow sediments match some of the subglacial and supraglacial tills very closely. Distinctions between sediment types using only particle size criteria cannot be sustained because the particle size characteristics of tills and debris flow sediments are remarkably similar. Due to the widespread re-working of glaciogenic sediments, grain shape and sand grain surface textures are not diagnostic of depositional environment. The sedimentology of diamictons depends on slope variables, particle size range, water content and petrology. As a result of these variables a large variety of sedimentary structures may be produced. However, facies associations are important in helping to distinguish diamictons, although it must be recognised that debris flows may occur in glacial and glaciolacustrine environments and the possibility of reworking may lead to misinterpretation.

A wide range of sedimentological and structural features have been recognised in the field. Debris flows range from gently inclined to steeply inclined flows, and are usually leveed, while debris slides are often bounded by lateral and basal shear zones. Glacial sediments range in origin from meltout to lodgement and slide facies, each with characteristic fabrics and

10

sedimentary structures related closely to depositional slope, particle size distribution and water content.

Great care must be taken in differentiating the various diamictons found within the Karakoram and western Himalayas, as well as in similar high mountain regions. This is very important when reconstructing palaeoenvironments and glacial histories. The identification of the various diamictons requires a multivariate approach. This paper will describe the sedimentological and geomorphological variation of contemporary and ancient diamictons from the Karakoram Mountains and western Himalayas as a basis for eludicating the genesis of diamictons of problematic origin in the Himalayas and adjacent regions. Emphasis is place on the use of micromorphology as an aid to providing new information on the nature of till and mass movement sediments in this region.

2 THE GLACIAL SYSTEM

A detailed discription of the glacial system in the Karakoram Mountains has been provided by Owen and Derbyshire (1989). They stressed the importance of meltout and supraglacial processes, and the complex interaction between the glacial, paraglacial, proglacial and periglacial process environments.

The ice masses in the Karakoram Mountains and the western Himalayas are glaciologically complex: their high altitude source areas have permafrost and annual precipitation totals in excess of 2000mm, while their snouts extend down to semi-arid valley floors. Mean annual temperatures at the equilibrium lines are generally below $0^{\circ}C$. As a result of their great thickness, however (often exceeding 400m), basal sliding is important, and the combination of high accumulation rates in the upper reaches and high ablation rates in the lower reaches in summer induces high values of mass flux, glacial abrasion, glacial debris transport, meltwater runoff and glaciofluvial sedimentation. The steep, unstable valley sides both of rock and older glacial deposits provide large volumes of debris, often mixed with snow which avalanche on to the glacier surface. The supraglacial and englacial sediment flow paths are thus dominant at present, the basal debris zone being relatively thin. The tills currently being deposited, therefore, are mainly of the meltout type with an important input from supraglacial sliding. Adjacent to the unvegetated valley sides, re-radiated solar energy values are high: this isolates very large, lateral moraines with varying amounts of ice core. In the lower reaches, lakes occur on glacier surfaces and at ice fronts (both glacier dammed and moraine dammed). The

moraines show wide variation in facies from laminated silts to coarse bouldery diamictons.

Owen and Derbyshire (1988 and 1989) recognised two major types of glacier based on the landform associations in their lower reaches. These were named after the type glacier (Figure 2). The first of these, the "Pasu type" is characterised by hummocky moraines and glaciofluvial outwash plains (Figure 2B). Small end moraines are present comprising hummocks and parallel ridges. Thin lodgement tills (<3m thick) are also present often plastered against bedrock knolls. Tills are easily identified in this setting and debris flows are less common. The second or "Ghulkin type", comprises large end moraines and ice-contact fans summounting the glacier's snout and reaching heights of several hundred metres (Figure 2C). Here supraglacial meltout tills, flow tills, debris flows, and glaciofluvial sediments, as well as tills resedimented by debris flow and slide processes are present. These are interbedded and intercalated forming complex assemblages of sediment which, when dissected by outwash streams, provide good sections. Figure 3 is a schematic diagram incorporating the landforms and sediment types present in the glacial and paraglacial environment in the Karakoram Mountains. There are several important landforms associated with the Karakoram and Himalayan glaciers, these will be described in detail below.

2.1 Lateral moraines and "Ablation valleys"

Large lateral moraines are well formed along most glaciers in the Karakoram and Himalaya, with reliefs which frequently exceed 20m above the ice and more tnan 100m above the valley floor (Figure 2C). Two types of lateral moraines have been recognised (Owen and Derbyshire, 1989). The first forms steep ridges comprising poorly sorted tills with diffuse fabrics. Clasts are orientated with their a-b planes subparallel to the slope of the moraine and their a-axis dipping away from the valley axis, transverse to the trend of the moraine ridge. These moraines often form nested complexes, the ridges becoming progressively larger nearer to the present ice margin. Ice-cored moraine is common near the present day ice margins, where meltwaters can be seen to be escaping from the moraines. These meltwaters help initiate sliding and collapse of large sections of the moraine. Slurrying of till due to waters escaping from the moraine flushes fines onto the moraine surfaces forming silty coats on top of the coaser diamictons. These, in turn, may eventually become covered by sliding of coarser material and are incorporated into the

Figure 2. Photographs showing the variations in contemporary glacial environment and glacier types in the Karakoram mountains. A. View of the Batura glaciers, saturated with supraglacial debris. Note the differential rock varnishes giving progressively darker tones with age picking out three phases of hummocky moraine development. The glacier is about 1km wide; B. View of the Pasu glacier, note the lateral moraine (right), the glacially scoured rock (centre) and the proglacial lake (foreground). Also note the glacier ice has little supraglacial debris on its surface compared to Plate 3A, C & D. the glacier is about 600m wide; C. View of the Ghulkin glacier showing well developed lateroterminal and end moraines, and outwash fans. Note houses and terraces for scale; D. View of the Gulmet end moraine with a glaciofluvial outwash fan developing between moraine ridges.The largest boulders in the foreground are about 2m in diameter.

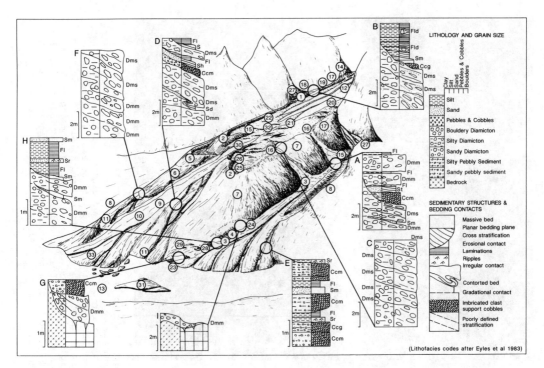

Figure 3. Landforms and sediment associations of ice-contact facies based on a variety of glaciers and glacier types inthe Karakoram Mountains and western Himalayas. 1. truncated scree; 2. lateroterminal dump moraine; 3. outwash channel draining laterally; 4. glaciofluvial outwash fan; 5. slide moraine; 6. slide-debris flow cones; 7. slide modified lateral moraine; 8. abandoned lateral outwash fan; 9. meltwater fan; 10. meltwater fan; 11. abandoned meltwater fan; 12. bare ice areas; 13. trunk valley river; 14. scree slopes being resedimented by debris flow processes; 15. supraglacial moraine; 16. gullied lateral moraine; 17.lateral moraine; 18. ablation valley lake;19. ablation valley; 20. supraglacial lake; 21. poorly developed medial moraine; 22. ice thrusts; 23. roches moutonnées; 24. glaciofluvial terraces; 25. fines washed from supraglacial moraines; 26. ice cored moraines; 27. scree; 28. ice scoured rocks; 29 hummocky moraines; 30. gullied frontal moraine; 31. island of eroded glaciofluvial terraces; 32. supraglacial debris; 33. glaciofluvial outwash plain. Sedimentary logs: A. Meltout tills from a lateral moraines overlain by glaciofluvial and debris flow sediments; B. Meltout and slide tills from a lateral moraines overlain by ablation valley lacustine sediments; C. Meltout and slide tills of a lateral moraine; D. Meltout and slide tills, and glaciofluvial sediments of the end moraine complex; E. Glaciofluvial outwash sediments; F. Meltout, slumped and slide tills of the end moraine; G. Lodgement till and glaciofluvial outwash on a roches moutonnées; H. Meltout tills and lacustrine sediments of a hummocky moraine; I. Lodgement till on bedrock.

moraine as thin sand/silt lenses. On a larger scale (1-10m) a crude stratification dipping down valley sub-parallel to the present-day ice gradients is observed where sections through ancient moraines are available or where ice has down wasted revealing the exposed and unsupported slopes of the moraines. These moraines probably represent successive phases of moraine development as ice thickens and wastes, depositing supraglacial moraine (cf. Osborn, 1978).

The second type of lateral moraine comprises overconsolidated tills which are plastered on to the valley walls at steep angles (> 40°). Clasts are oriented in a similar manner as the first type of lateral moraine. Sheets of till break away parallel to the slopes due to stress release mechanisms. Owen and Derbyshire (1989) believe these to be produced subglacially by lodgement processes along the margins of the ice during times of extreme ice thickening.

Frequently, a valley is present between the lateral moraine and valley wall. These have been described by Mason (1929) as "ablation valleys", although this term is

Figure 4. Photographs showing the variability of structures within diamictons in the Karakoram Mountains. A. Hummocky moraines near the snout of the Batura glaciers. Note deformation structures in the centre of the frame picked out by a silt-rich intercalation. These may result from a combination of slumping during ice-core melting and glacial push. Also note very angular poorly sorted clasts; B. Subglacial lodgement tills on bedrock surfaces near the snout of the Pasu glaciers. Note edge-rounded clasts and glacially smoothed bedrock. The field of view is about 12m; C. Dissected end moraines at the snout the Ghulkin glacier. Note steeply angled stratification dipping towards the right of the frame. This may be the result of deposition on steep slopes and the resultant slump and slide processes during and after deposition. The section is about 20m high from the bottom to the top of the frame; D. Tills resedimented by debris flow processes forming terraces near Gilgit. Note low angle stratification and the similar lithology to that of the till sediments in Plate 6A & C. The terrace is about 5m high.

14

Figure 5. Details of diamicton lithologies. A. Resedimented till sediments in a terrace section near Gilgit. Note diffuse isotropic fabric. B. Subglacial lodgement tills from Skardu, note compact fabric and edge-rounded clasts; C. Meltout till from near the Batura glacier, note the similarity with Figure 7A; D. Hummocky moraine from near the snout of the Batura glacier. Note the dominance of silt in this till and the deformed stratification.

inappropriate as these valleys are not formed by ablation processes, rather they act as zones for the accummulation of snow and sediment. Deposits within these valleys include mass movement, lacustrine, glaciofluvial and till sediments. The lateral moraines are commonly deeply dissected allowing melt waters to flow from the ice into the ablation valley. The collapse of moraine walls is common adjacent to these stream cuts and the channels often act as paths down which water saturated in glacial debris and debris flows advances from the glacier into the valley.

2.2 Hummocky moraines

Hummocky moraines are common in all the glacial environments in the Karakoram and Himalayan mountains, and comprise a chaotic assemblage of depressions and hummocks that may reach several tens of metres in height (Figure 2A & B). The hummocks are composed of supraglacial meltout till interstratified with glaciolustrine and glaciofluvial

sediments. Contemporary hummocky moraines often have small ponds within depressions. Streams drain into them as a consequence of surface flow and seepage from melting ice-cored moraines. These hollows quickly infill with sand and silt. Till fabrics are diffuse and clast fabrics within hummocky moraines have a low degree of preferred orientation. Small (cm-wide) pipes rich in silts represent water/ sediment escape structures produced by grain size differentation as the ice core melts and waters seep to the surface. Laminations of sand and silt produced by supraglacial waters are present and are often distorted by slump processes due to ice core melt and hummock collapse (Figures 4A, 5C and D). Ice-push structures are also present in many of the hummocky moraines which suggest that a simple meltout process is not adequate to describe all the structures present within the moraines.

15

2.3 Subglacial moraines and roches moutonnées

Subglacial tills are rare, and when present infill depressions and form small hummocks and flutes. These are often obscured by a large quantity of supreglacial tills and glaciofluvial sediments. Subglacial lodgement tills, where present, comprise sheared, overconsolidated, silt, rich diamictons with strongly orientated clast fabrics (Figure 5B). Clasts are striated, edge-rounded and bullet shaped (Figure 4B). These are often found on the sides of bedrock knolls forming roches moutonnées. Subglacial meltout tills are rare and where present represent very young deposits which are being actively eroded: they have poor preservation potential. These comprise silt-rich diamictons with a diffuse fabric, and poor stratification usually comprising sandy silts lenses or sheared sediment.

2.4 End moraines and ice-contact fans

These landforms are cone shaped or arcuate in form comprising till and glaciofluvial sediments which are deposited on steep slopes (c. 15-25°) around the snout of the glacier (Figure 2C & D). Debris-flow and slide processes are commonly initiated by melting ice cores and glaciofluvial streams. This produces resedimented supraglacial tills interstratified with glaciofluvial sediments (Figure 2D & 4C). Tills exhibit crude, metre thick, stratification parallel to the depositional slope (Figure 2C). Such tills are poorly sorted, with diffuse fabrics which are weakly orientated with clast a-b planes subparallel to the stratification. Successive arcuate end moraines are progressively infilled with glaciofluvial sediments as meltwater channels aggrade and dissect the morainic ridges (Figure 4D). Some small end moraines are asymmetrical and exhibit a bimodal fabric with strong up valley dips relating to glacial push and down valley dipping fabrics relating to sliding of till after deposition by slope processes.

3 SLOPE PROCESSES

Mass movements are important and frequent in the Karakoram and Himalayan mountains, acting on the very long steep slopes. Processes include: rockfalls, avalanches, debris slides, rockslides, creep, rotational slides, debris flows and flowslides (Owen, 1989 & 1991). Events vary in scale from small isolated failures to extensive areas of complex movements. Many of the mass movement deposits originate from previously deposited glacial and mass movement sediments. Frequently,

the final sediment is polygenetic in origin and may have been reworked several times by different mass movement processes (Owen, 1991). Particularly important processes include large scale debris flows and debris slides. These commonly occur adjacent to recent lateral and end moraines. Derbyshire and Owen (1990) suggested that much of the till deposited by past glacial advances has been resedimented by these mass movement processes to form impressive fans in the valley floors, which now have alluvial fan-like geometries. Large-scale failure within confined valleys produces steep (15-30°) cone-shaped fans, often damming the river to form short-lived lakes.

3.1 Rockfalls and rock avalanches

Rockfalls range from small scale toppling failures to large debris falls. These originate from bedrock slopes or deeply truncated or dissected moraines. Falls are commonly initiated by melting snows, frost shattering, heavy rainfall, snow and ice avalanches, earthquakes, erosion and sediment loading. Rockfalls and rock avalanch material accounts for the majority of supraglacial debris and debris within ablation valleys, where they form large block fields comprising dominately angular boulders. Rockfall material, however, does not only produce coarse grained material but large quantities of silt size debris are also produced as rocks disintegrate on impact. Some rockfalls may even result in sturzstroms as material disintegrates and is carried down valley by flowslide and grain flow processes (cf. Hsu, 1975). Rockfall deposits usually comprise angular, very poorly sorted, randomly orientated, clast supported sediments. Rock and snow avalanches produce similar deposits to rockfalls, but avalanche deposits can easily be distinguished by the presence of small mounds and pyramids of debris which stand proud on many of the boulders, left after the avalanche snow has melted.

3.2 Debris flows

Classic examples of debris flows can be recognised thoughout the Karakoram Mountains. Debris flow material is frequently derived from moraines and debris are channelled down well defined leveed channels leading to accumulation areas, where material spreads out via distributaries. This is very characterisic of a Bingham's style plastic flow (Johnson & Rodine, 1984). Many debris flows are initiated in the spring and summer as snow and glacial ice melts allowing meltwater to excavate or saturate sediments such as glacial deposits or previously deposited mass movement sediments.

3.3 Debris slides

Sliding of debris is an important process in many of the contemporary glacial environments. Sliding may occur on a small scale as individual grains slide down the surface of landforms as meltwaters from an ice core or melting snow helps lubricate the surface. This sliding is probably the major process responsible for the crude stratification seen in many of the lateral and end moraines. Debris slides can also be recognised within till sediments and on a large scale (in the order of several metres). These slides are gently inclined ($<10°$) to steeply inclined ($>45°$) with basal shear surfaces and subvertical lateral shears. Movement is accommodated along these shear zones. Shear zones are more commonly seen in ancient diamictons than those being actively deposited today. Slickensides are common on these surfaces, resembling the striated surfaces found in lodgement tills. In debris slides, however, the shear surfaces dip down valley, whereas in tills shear surfaces dip up valley or up ice movement direction.

Many of the mass movement deposits do not show these diagnostic morphological features because they are quickly eroded by fluvial or other mass movement processes, or become buried beneath thick valley fills. When this is the case detailed sedimentological criteria must be used to identify the deposits.

4 SEDIMENTOLOGICAL CHARACTERISTICS OF GLACIAL AND NON-GLACIAL DIAMICTONS

The sedimentological characteristics of diamicton deposits reflect the source material, the climatic regime and the processes of transportation and deposition. The first of these includes the lithology at the glacier bed, the rock slopes above the glacier and and in the case of diamictons which have been resedimented the nature of the sediment, its lithology, and geotechnical and physical properties. Lithology determines the degree of weathering and erosion, the types of secondary minerals which are likely to be produced, which in turn helps to determine the physical and geotechnical properties of the sediments, plus the size and shape characteristics of its clasts. The climatic regime influences the availability of water (the rates of precipitation and melting of ice), the rate of weathering and the rates and types of sediment transfer processes, especially glaciological and mass movement regimes. Processes of transport and deposition determine facies associations, particle size and shape, fabric, sedimentary structures, geotechnical properties, micromorphology and distribution of the sediments. Each set of

sedimentary characteristics will be discussed in detail.

4.1 Facies Associations

Figure 3 illustrates the range of facies associations present around the active margins of a characteristic glacier in the Karakoram and Himalayan mountains. Taken in context with morphologies and associated landforms most units can be readily distinguished and processes of deposition can be assigned to them. Taken out of context, however, as is the case with many ancient dissected landforms, or where the landforms are adjacent to the valley walls which were previously glaciated, it is difficult to distinguish landform types. Sections such as A, C, D, F and H in Figure 3 illustrate how till sediments may be confused with mass movement deposits if facies associations are considered in isolation. Sections such as B which contains lacustrine silts with dropstones (Fld) clearly indicate the proximity to ice. The glacially smoothened bedrock surfaces with till squeezed into joints, such as sections G and I unequivocally shows that deposition was by glacial lodgement processes. In the majority of cases, however, elucidation of the genesis of diamictons can not be made on the basis of facies analysis alone.

4.2 Particle size distributions

Particle size distributions in tills are a function of many factors which include comminution of rock by glacial erosion; processes of transportation; and the mode of deposition. In the Karakoram Mountains, the supraglacial environment dominates the depositional system and when sediments are incorporated into the glacier they often pass through the ice via passive flowlines resulting in little modification of the sediment. However englacial sediments show some degree of modification during deposition by processes of illuviation and mass movement as a result of the abundant meltwaters. Particle size characteristics are therefore dominated by the processes by which material falls onto the glacial ice and by the mode of deposition, which in turn is controlled by the abundance of meltwater and the slope charcteristics. The particle size distributions of all supraglacial tills are very similar: very poorly to poorly sorted, coarsely skewed and depleted of fines ($<10\%$: Figure 6 & Table 1). Subtle differences, however, can be recognised with tills of different lithologies (Owen and Derbyshire, 1989). Tills derived from limestone source areas, for example, are finer grained than those derived from granitic source areas.

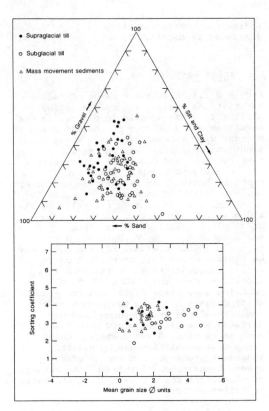

Figure 6. Ternary plot and co-plots illustrating the particle size variability of diamicton sediments from the Karakoram Mountains and western Himalayas.

Subglacial tills are similarly very poorly to poorly sorted, positively skewed, but have finer mean grain sizes than the supraglacial and englacial tills.

Particle size characteristics in mass movement deposits are controlled by the particle size distributions of the source material, and the mode of transportation and deposition. The majority of the source material that is moved by slope processes is primarily derived from glacial sediments. Therefore, particle size distributions are very similar to those of till sediments. Little modification of the particle size distribution takes place during the resedimentation of tills by mass movement processes. The particle size distributions are very poorly to poorly sorted, coarsely skewed, but are slightly finer grained than subglacial tills. Co-plots of supraglacial tills and mass movement deposits further demonstrates the broad similarities between the two types of sediment, but they clearly demonstrate the differences between supraglacial and subglacial sediments. Similar characteristics were recognised by Lawson (1979) for resedimented tills/flow

Table 1. Particle size data for tills and debris flows.

Sample	Mean (phi)	Sort- ing	Kurt- osis	Skew- ness
Debris flows	2.25	4.00	0.35	-0.57
	1.76	3.26	0.59	-0.35
	1.27	3.11	0.62	-0.75
	2.13	3.03	0.79	-0.12
	1.96	2.85	1.49	-0.28
	1.28	3.37	1.05	-0.35
	1.66	3.86	0.51	-0.25
Debris slide	2.02	3.14	0.63	-0.16
Hummocky moriane	2.83	4.05	0.42	0.09
	1.51	3.60	0.65	0.74
	1.35	2.97	0.38	-0.29
	2.03	2.50	0.63	-0.47
Lodgement till	1.62	4.12	0.43	0.14
	2.45	3.16	0.51	0.10
	2.89	3.34	0.81	0.21

tills at the terminus of the Matanuska Glacier in Alaska.

4.3 Particle shape

Particle angularity and roundness is dependant on lithology and mode of transportation. In the Karakoram and Himalayan glacial systems till clasts undergo very little modification because supraglacial debris and debris entrainment via passive flow paths dominate. However, boulders, cobbles and pebbles exhibit a low degree of edge rounding, but striations are rare. Finer grain sizes, however, remain very angular. Scanning electron and optical microscopy studies of subglacial tills show clear edge-crushing and fracturing of silt and sand grains (Figure 7 to 11, especially Figures 7B & D, & 8D). This, however, does not result in a reduction of the angularity of these fine particles, rather it is an important process in producing very angular silt size particles. Roundness is controlled mainly by lithology. This is very noticable when comparing oblate clasts in tills derived from granitic source area with the tabular clasts derived from slate rich source areas (cf. Figure 9B and 9C).

Particle angularity and roundness characteristics in mass movement sediments reflect the source material. These can range from angular blocks deposited by rockfall from rock slopes to edge rounded clasts when the material has been derived from tills.

4.4 Fabric

Fabric describes the directional properties of sediments which include stratification and bedding, water escape structures, pebble fabric, compaction, jointing, shearing and pebble fabrics. Fabrics can be observed on all scales from the microscopic to several tens of metres. They may be primary, produced during deposition, or secondary, produced by post depositional processes.

4.4.1 Stratification and bedding

Figure 3 illustrates the characteristic stratification and bedding within tills and mass movement deposits that can be seen in the field. These vary from abrupt (Figure 4D) to gradational contacts to crude stratification picked out by zones of more open frabrics (Figure 4A & C). The latter characteristics are very common in supraglacial and mass movement deposits in the Karakoram Mountains. These represent almost continous deposition on slopes with fluctuating glacial meltwater and snow melt discharges. This influences the type of flow and slide process. Grain size variations may also be produced by changing flow regimes due to fluctuating meltwater discharges. Where finer grained intercalations are present within the diamictons these are commonly deformed into slump or open irregular folds (Figure 5D). This illustrates the importance of slump and slide processes. Sharp bedding contacts often represent abrupt changes in sedimentary environment frequently produced by changing positions of supraglacial or subglacial meltwater streams.

4.4.2 Water-sediment escape structures

These are injections of fine sediment within the diamictons. They vary in size from the microcsopic scale to several cm in diameter and are usually composed of finer grain sizes than the surrounding matrix. (Figure 12). The a-axis of clasts and grains within these structures become aligned parallel to the lenght of the structure. These features are most common in hummocky moraine which have formed by melting of ice-cores, similar to those described by van der Meer (1987) in the Alps. However, they are also present in subglacial tills, possibly produced as water and sediment is squeezed out by ice and/or lithostatic pressures, or as interstratified ice melts.

4.4.3 Pebble fabric

The use of a-b plane pebble fabrics provides important information on the stress field and flow regime involved in the deposition of tills and mass movement deposits. In simple terms the clasts' a-b planes dip up valley in sediments derived from glacial shearing processes, such as lodgement tills and subglacial meltout

tills. Whereas supraglacial tills and sediments deposited by mass movement processes dip down valley. Complexities occur, however, when post-depositional modifications reorientate clasts. In addition, some mass movement mechanisms produce more complex fabrics. This is the case with debris flows characterised by Bingham plastic flow. The clasts' a-b planes align themselves parallel to the channel sides and levees. End moraines may also show bimodal fabrics, one dipping up valley as sediment is sheared into position by glacial processes, the other down valley on the lee side of the moraine as material slides down the front of the glacier. In summary, pebble fabrics can provide one of the most useful and important criteria to help resolve the genesis of Karakoram diamictons.

4.4.4 Compaction

The degree of compaction can be quantified in sediments of similar lithologies by measuring their bulk densities (Akroyd, 1969; McGown and Derbyshire, 1977). Compaction can also be examined using scanning electron and optical microscopy. Compaction is a function of glacial and lithostatic loading and the desiccation of the sediment. Sediments which have been buried under great thicknesses (> 100m) of sediment begin to demonstrate some degree of lithostatic consolidation. Disregarding burial, lodgement tills show a marked increase in bulk density compared to supraglacial tills and mass movement deposits (Table 2). Microscopically the degree of compaction may vary considerably throughout a single sample (Figure 9A & B). A meltout till, for example, will generally have a diffuse fabric, but there may be zones where local dewatering has produced consolidation by desiccation. In contrast, lodgement tills generally have densely compacted fabrics, but zones of open fabric can be recognised produced by dilation during shearing.

Compaction may be altered by secondary processes, such as illuviation of fines and

Table 2. Bulk densities of tills and debris flow sediments (gcm^{-3}). After Owen and Derbyshire (1989)

Lodgement till	2.45
	2.41
Debris slide	2.39
	2.38
Massive till	2.22
Lodgement till	2.17
Debris flow	2.17
	2.08
Hummocky moraine	2.08
Debris flow	2.04
Hummocky moraine	1.97

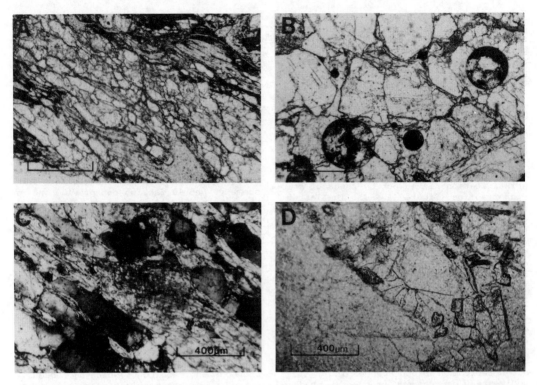

Figure 7. Photomicrographs of vertically orientated thin sections of tills, all are viewed in plane polarised light and the scale bar is 4mm unless otherwise stated. A. Till form a lateral moraine from Shatial showing shear structures indicating movement from right to left; B. Lodgement till from Shatial with a tightly packed fabric, with a strong grain contacts; C. Enlarged view of the shear zone in Figure 9A showing bending and shearing of micas; D. Lateral moraines from Shatial showing an impact fracture between quartz grains.

dissolution creating secondary porosities (Figure 10F). In addition, cementation may reduce porosity and effectively increase the bulk density of the sediment. This is important because bulk densities are not meaningful when comparing sediments unless their micromorphology has been examined.

4.4.5 Jointing

These relate to depositional and post-depositional stresses, commonly the result of stress release mechanisms or desiccation as sediments become consolidated as they dry out. Several types are recognised including sub-horizontal, sub-vertical and parallel to slope. These range in scale from diffuse microscopic joints to pervasive macro joints with lengths in the order of several metres. All types are common in overconsolidated sediments such as lodgement tills, but are less common in meltout tills and mass movement deposits. On a microscopic scale platey minerals may align themselves parallel to these joints

orientated by the palaeostress field.

4.4.6 Shearing

Shear systems in diamictons vary in scale from the microscopic to bands of shears measuring several cm across (Figure 7A & C, 8F, & 10A, B & D). They may consist of simple zones of shear to complex anastomosing systems. Microscopic shearing is described in detail in the section describing the micromorphology. On a macroscale shears within lodgement tills dip up ice movement direction while in debris slides they dip down movement direction. Commonly, shears are slickensided (Figure 11) and strain hardened and in tills they often form positive relief as the adjacent sediment is more easily eroded away. In debris slides, shears are rarely strain hardened and may be lined with minerals such as kaolinite, though they may become cemented at a later date which increases their strength.

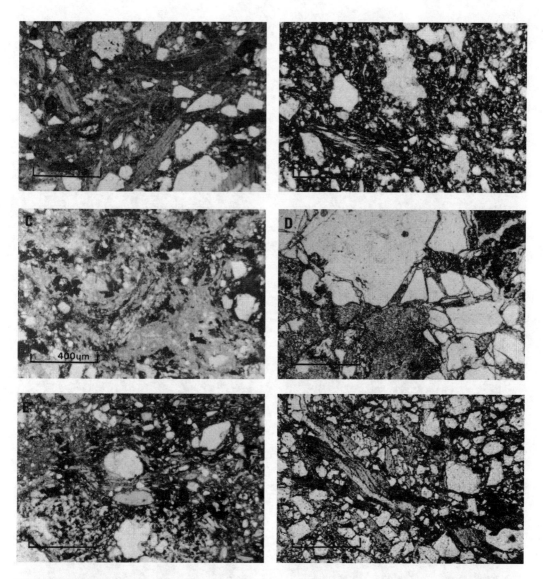

Figure 8. Photomicrographs of vertically orientated thin sections of tills, all are viewed in plane polarised light and the scale bar is 4mm unless otherwise stated. A. Till resedimented by debris flow processes from the Bagrot valley; B. Debris flow sediment from Batkor; C. & E. Debris flow sediment from near Gilgit, note the swirly structures picked out by the fine clays; D. Till resedimented by debris flow processes from near Gilgit, note the fractured clast which can be easily pieced together; F. Debris slide sediment from near Batkor, note the discrete shear system.

4.5 Micromorphology

Micromorphology, involving scanning electron and optical microscopy, is particularly important in providing additional information regarding the sedimentology of the diamicton. Sadly, little use is made of these techniques which are well described by Derbyshire (1978) and van der Meer (1987). Figure 12 is a schematic diagram illustrating the typical structures that can be recognised in a diamicton from the Karakoram and Himalayan mountains using both optical and scanning electron microscopy. It is important to note that similar structures can be identified in each type of sediment. Shears, for example, can be identified in

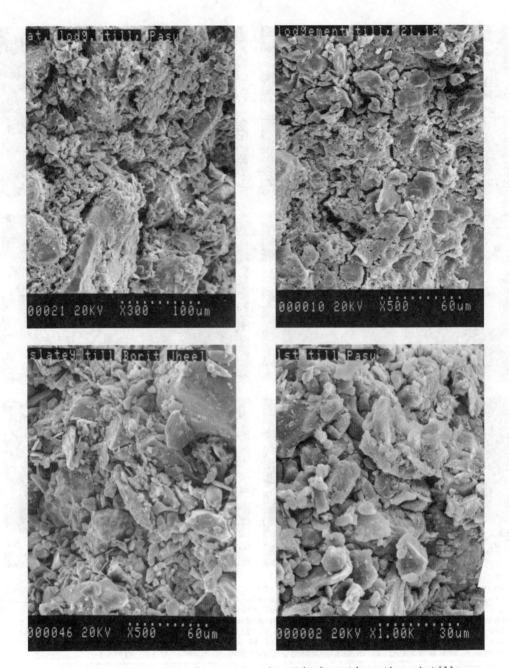

Figure 9. Scanning electron photomicrographs of vertical sections through tills.
A. Lodgement till from a lateral moraine near the Pasu glacier. Note the areas of
differential overconsolidation and areas of high porosity; B. Lodgement till from a
bedrock surface in the Gilgit valley. Note the compact matrix and irregular dilation
cracks due to overconsolidation; C. Slate-rich till from a lateral moraine at Borit
Jheel. A weak anisotropy has developed with elongated clast dipping towards the right
indicating an ice movement direction towards the left. Note dilatancy due to the
shearing of platey grains; D. Limestone-rich subglacial meltout tillfrom near Pasu.
Note the granular porous matrix without and ordered structure.

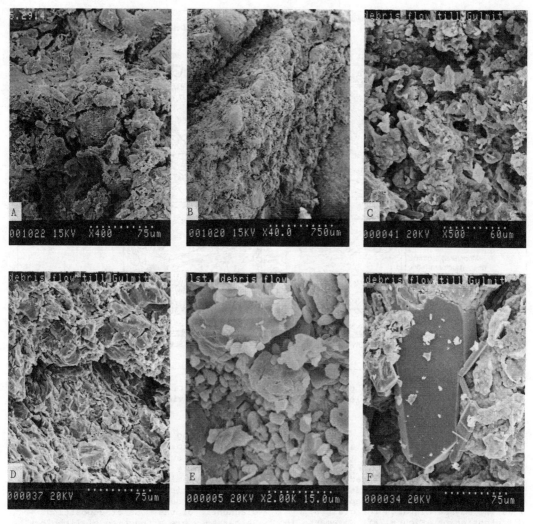

Figure 10. Scanning electron photomicrographs of vertical sections through mass movement sediments. A. & B. Stratified debris flow sediment from near Gilgit. Diffuse fabric and low angled zones of discontinuity picked out by platey silts (upper part of A. and upper left in B.) Flow direction out of page in A. and right to left obliquely out of page in B; C. & D. Till resedimented by debris flow processes from the valley floor near Gulmet. Zone of compact fabric cemented with carbonate cement. A porous fabric is evident where there is little cement. many of the clasts are fractured but their constituent parts are not widely separated; E. Debris flow sediment from near Shatial. Small euhedral crystals are frequent. these are probably feldspar (EDX analysis). They remain intact within the flow but would have probably been greatly abraded within a glacial sediment; F. Till resedimented by debris flow processes from near Gulmet. Euhedral calcite crystals and a fine acciular cement within a secondary pore.

lodgement till and debris slide sediments, and the alignment of clays due to dewatering in meltout tills may be misidentified as shears (pseudoshears). Distinctions between the types of structures can be made in terms of their orientation in a vertical plane with respect to ice movement direction. In lodgement tills the dominant shear systems dip up ice movement direction whereas in debris slides or resedimented tills shearing dips down ice movement direction or down valley. In meltout tills, pseudoshears are irregular in orientation

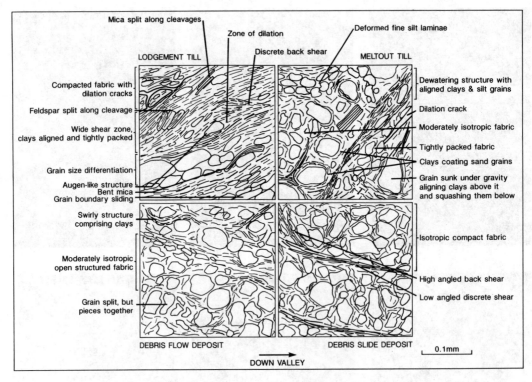

Figure 11. Scanning electron photomicrographs a lodgement till showing a striated surface, movement direction left to right. Total field of view is about 4mm.

and are often subvertical.

A crude stratification can also be recognised in all sediments. In lodgement tills this is tectonically induced as larger grains become entrained in zones of higher stress (a pseudolamination). Similarly, in debris slide deposits the shearing may be mistaken for grain size differentiation (Figure 7A). Movements in debris flow sediments may also produce a similar effect (Figure 8F), but they frequently produce "swirly" structures (Figure 8C & E). In meltout tills true laminations are common, but unlike the psuedolaminations in lodgement tills and debris slides they are usually deformed, a consquence of the irregular settling of the sediment during deposition as the ice-core melts.

The degree of packing also varies between types of deposits, from compact edge to edge packing to diffuse open fabrics (cf. 7B with 8B). Within tighly compacted fabrics such as lodgement tills there are often zones of open fabric produced by dilation due to irregular shearing (Figure 9C). Whereas in meltout tills which generally have an open diffuse fabric zones, tightly compact fabric can often be recognised. These compact fabrics are frequently associated with dilation cracks

and dewatering structures which suggests they result form desiccation.

Sand grain surface textures are not discussed here in detail because they are of little use in studying sediments which are polygenetic in origin. Particle shape in silts and sands are generally not diagnostic of the mode of deposition. Rather they are a function of lithology, the quartz and feldspar grains derived from granitic rocks, for example, are very equant whereas all grains from slates and schists are generally elongate and tabular. Silt grains are very angular in all sediments (Figures 7 to 10). In lodgement tills, however, edge crushing is frequently observed in silt grains and fracturing of grains by the stress induced by the impact of other grains is common (Figure 7B). Larger grains are frequently seen to have been broken up and strung out in lodgement tills often forming augen like structures (Figure 7A). These features are not common in meltout tills, debris slide and debris flow sediments. In debris flow and debris slide sediments, however, it is possible to see larger grains broken up and in a jigsaw-like configuration so that they can easily be pieced together (Figure 8D). These grains are not strung out and do not form augen-like structures, rather they probably

Figure 12. Diagram summarising the characteristic micromorphologies that can be recognised in diamictons of different origin within the Karakoram Mountains. The diagram is based on thin section work and scanning electron microscopy.

Table 3. Mineralogy of the silt and clay fractions for selected tills and mass movement sediments

Mica	Chlorite-vermiculite	Kaolinite	Illite	Chlorite	Smectite	Dickite
Debris flows						
Saturated	minor	minor	major	minor	---	---
Saturated	minor	minor	major	trace	---	---
Saturated	trace	trace	moderate	---	---	---
Saturated	moderate	major	major	minor	---	---
Saturated	minor	moderate	major	trace	---	---
Tills						
Saturated	-----	moderate	trace	minor	---	---
Saturated	moderate	major	minor	major	---	---
Saturated	trace	moderate	---	---	---	---
Saturated	trace	major	major	minor	---	---

move passively together within the sediment as it flows.

4.6 Clay and silt mineralogy

The majority of tills and mass movement deposits in the Karakoram mountains have <10% clay-size fraction and often as little as 5%. The silt and clay fractions are therefore considered together. Table 3 lists the silt and clay mineralogy determined by x-ray diffraction for tills and mass movement sediments from the Karakoram and western Himalayas. The mineralogy is essentially detrital in origin with a dominance of quartz, feldspars, kaolinite and micas derived from

granitic rocks and chlorites and micas derived from metamorphic rocks. No "active" clay minerals were identified such as swelling clays. Complex chemical reactions and extreme volume changes involving these minerals are not considered an important process in the western Himalayas. The presence of much Kaolinite, however, may aid sliding processes as this mineral acts as an important lubricant.

5. CONCLUSION

Table 4 summarises the main types of sediments and their properties and characteristics. The distinctions between the various diamictons is complex, and

25

Table 4. The dominant sedimentary characteristics of diamictons of different genetic origin in the Karakoram Mountains and western Himalayas.

Origin	Sedimentary structures	Clast fabric	Clast angularity	Particle size characteristic Mean/phi	Sorting	Skewness	Mineralogy	Bulk density	Micromorphology
Subglacial lodgement till.									
Subglacial lodgement processes	Sheared; overconsolidation structures	Strongly orientated up valley	Edge rounding; angular	Fine sand-coarse silt	Poor	Fine	No active clays	High	Shears; overconsolidation; high grain anisotropy; edge crushin
Subglacial & englacial meltout till									
Subglacial meltout & shearing	Massive; sandy lenses; shears	Low anisotropy	Slight edge rounding	Med. sand-coarse silt	Very poor	Coarse	"	Low	Diffuse fabric; shears; dewatering structures; laminae
Supraglacial meltout till									
Deposition of subaerial till as ice melts out	Massive; slumped; intercalations; dewatering struct.	Low anisotropy	Very angular-angular	Med. sand-coarse silt	Very poor-poor	Coarse	"	Low	Diffuse fabric; dewatering structures; microlaminae
Supraglacial slide till									
Slide of till down ice or moraines	Massive; crude down-slope stratification	Low anisotropy	Very angular-angular	Med. sand-coarse silt	Very poor	Coarse	"	Low-medium	Diffuse fabric; dewatering structures; shears
Supraglacial flowtills									
Flow of wet till down ice or moraines	Massive; downslope stratification; leveed channels	Weak down-valley	Very angular-angular	Med. sand-coarse silt	Very poor-poor	Coarse	"	Low-medium	Diffuse fabric; dewatering structures; shears; moderate clast anisotropy
Debris flow sediments									
Debris flow of weathered rock or resediment-ation of diamicts	Massive; downslope stratification; leeved channels; shears	Weak down-valley	Very angular-angular	Med. sand-coarse silt	Very poor-poor	Coarse	"	Low-medium	Diffuse fabric; dewatering structures; shears; moderate clast anisotropy; swirly structures
Flowslide sediments									
Flowsliding of diamictons or rockfalls	Massive; crude sub-horizontal stratif-ication; shears	Low anisotropy	Very angular-angular	Med. sand-coarse silt	Very poor-poor	Coarse	"	Low-medium	Diffuse fabric; dewatering structures; swirly structures;
Rockfall sediments									
Rockfall	Massive; structureless	Isotropic	Very angular	----	Very poor	Coarse	- - -	- - - -	--------
Rockslide sediments									
Rocksliding of weathered rock or diamicts	Massive; shears	Very low anisotropy	Very angular-angular	Med. sand-coarse silt	Very poor	Coarse	No active clays	Medium	Compact-diffuse fabric; shears low clast anisotropy
Rock and snow avalanches									
Avalanching	Massive; earth pyramids	Isotropic	Very angular	----	Very poor	Coarse	- - -	--------	--------

there is a gradational series of sediments including lodgement tills, subglacial and supraglacial meltout tills, flowslides, debris flow, debris slide, creep and rock fall deposits. Frequently diamictons have been resedimented by one or more of these processes. Clearly, no one characteristic can be used unequivocally to elucidate the genesis of a particular diamicton in this active mountain environment. In addition, many of the characteristics and properties can be recognised in sediments deposited by all of these different mechanisms. An appreciation, however, of the processes of deposition and genesis of sediment, as well as the range of possible variations and overlaps between the different sediment types is imperitive in appreciating the genesis of these deposits. Such an appreciation is paramount in accurate reconstuctions of palaeoenvironments and the nature and extent of former glaciers.

ACKNOWLEDGEMENTS

Many thanks to D. Croots for critically reading the manuscript, and to the people of northern Pakistan for their hospitality, especially M.I.Khan (mountain guide and local supplier of exotic herbs and spices).

REFERENCES

Akroyd, T.N.W. 1969. Laboratory testing in soil engineering.-Soil Mechanics Ltd., London, p.437.

Brunsden, D. and Jones, D.K.C. 1984.The geomorphology of high magnitude-low frequency events in the Karakoram Mountains. In: Miller, K. (Ed.) . International Karakoram Project, Cambridge University Press, Cambridge, pp.536-579.

Derbyshire, E. 1978. A pilot study of till microfabric using the scanning electron microscope. In: Whalley, W.B. (Ed.): Scanning electron microscopy in the study of sediments, Geo Abstracts, Norwich, pp.41-61

Derbyshire, E. 1984. Sedimentological analysis of glacial and proglacial debris: a framework for the study of Karakoram glaciers. In: Miller, K. (Ed.) International Karakoram Project, Cambridge University Press, Cambridge, pp. 347-363.

Derbyshire, E., Li Jijun, Perrott, F.A., Xu Shuying & Waters, R.S. 1984. Quaternary glacial history of the Hunza valley, Karakoram Mountains, Pakistan. In: Miller, K. (Ed.) . International Karakoram Project, Cambridge University Press, Cambridge, pp. 456-495.

Derbyshire, E. & Owen, L.A. 1990. Quaternary alluvial fans in the Karakoram Mountains. In: Rachocki, A.H. & Church,M.

(Ed.): Alluvial Fans- a field approach, John Willey & Sons, Chichester, pp.27-55.

Dewey, J. F. & Burke, C.A. 1973.Tibetan, Variscan and Precambrian basement reactivation products of continental collision. J. Geol. 81: 683-92.

Le Fort, P. 1975. Himalayas- the collided Range. Amer. J. Sci., 275A: 1.

Eyles, N., Eyles, C.H. and Miall, A.D. 1983. Lithofacies types and vertical profile models; an alternative approach to the description and environmental interpretation of glacial diamict and diamictite sequences. Sedimentology, 30: 393-410.

Ferguson, R. 1984. Sediment load of the Hunza River.In: Miller, K. (Ed.) . International Karakoram Project, Cambridge University Press, Cambridge, pp. 374-382

Ferguson, R., Collins, D.N. & Whalley, W.B. 1984. Techniques for investigating meltwater runoff and erosion.In: Miller, K. (Ed.). International Karakoram Project, Cambridge University Press, Cambridge, pp. 374-382

Gansser, A. 1964. Geology of the Himalayas. Interscience, John Wiley, Chichester, p.289.

Goudie, A. 1984. Salt efflorescence and salt weathering in the Hunza Valley, Karakoram Mountains, Pakistan. In: Miller, K. (Ed.) . International Karakoram Project, Cambridge University Press, Cambridge, pp. 607-615

Goudie,A., Brunsden, D., Collins, D.N., Derbyshire, E., Ferguson, R.I., Hashnet, Z., Jones, D.K.C., Perrott, F.A., Said, M., Waters, R.S. & Whalley, W.B. 1984. The geomorphology of the Hunza Valley, Karakoram Mountains, Pakistan. In: Miller, K. (Ed.) . International Karakoram Project, Cambridge University Press, Cambridge, pp. 359-411.

Hewitt, K. 1968. The freez-thaw environment of the Karakoram Mountains. Canad. Geog., 12: 2, 85-98.

Hsu, K. 1975. Albert Heim: Observations on landslides and relevance to modern interpretations. In: Voight, B. (Ed.): Rockslides and avalanches, natural phenomena, Developments in Geotechnical Engineering, 14A, Elsevier Sci. Publ. Comp., Amsterdam, pp. 72-93.

Johnson, A.M. & Rodine, J.R.1984. Debris Flow. In Brunsden, D. & Prior, D.B. (Ed.): Slope Instability. John Wiley and Sons, Chichester, pp.257-362.

Lawson, D.E. 1979. Sedimentological analysis of the western terminus region of the Matanuska Glacier, Alaska. CRREL Rep., 79-9, 122pp.

Li Jijun, Derbyshire, E. and Xu Shuying 1984. Glacial and paraglacial sediments of the Hunza Valley, North-West Karakoram, Pakistan: A preliminary analysis.In: Miller, K. (Ed.) . International Karakoram Project, Cambridge University Press, Cambridge, pp. 496-535

Mason, K. 1929. Indus Floods and Shyok
Glaciers. Him. Jour., 1: 10-29.
McGown, A. & Derbyshire, E. 1977. Genetic
influences on the properties of tills.
Quat.J. Eng. Geol., 10: 389-410.
van der Meer, J.J.M. 1987. Micromorphology
of glacial sediments as a tool in
distinguishing genetic varieties of till.
In: Kujansuu, R. & Saarnisto, M. (Ed.):
INQUA Till Symposium. Geol. Surv.
Finland, Spec. Paper 3: 77-89.
Osborn, G.D. 1978. Fabric and origin of
lateral moraines, Bethartoli Glacier,
Garhwal Himalaya, India. J.Glac. 20: 84,
547-553
Owen, L.A. 1988a. Terraces, Uplift and
Climate, Karakoram Mountains, Northern
Pakistan. Unpublished PhD thesis,
University of Leicester, p.399.
Owen, L.A.,1988b. Wet-sediment deformation
of Quaternary and recent sediments in the
Skardu Basin, Karakoram Mountains,
Pakistan. In: Croots, D. (Ed.)
Glaciotectonics, A.A. Balkema, Rotterdam,
pp. 123-147.
Owen, L.A. 1989. Terraces, uplift and
climate in the Karakoram Mountains,
Northern pakistan: Karakoram intermontane
basin evolution. Zeit fur Geom., 76, 117-
147.
Owen, L.A. 1991. Mass movement depsoits in
the Karakoram Mountains. Zeit fur Geom,
35,4, 401-424.
Owen, L.A. & Derbyshire, E. 1988. Glacially
deformed diamictons in the Karakoram
Mountians, Northern Pakistan. In: Croots,
D. (Ed.) Glaciotectonics, A.A. Balkema,
Rotterdam, pp.149-176.
Owen, L.A. & Derbyshire, E. 1989. The
Karakoram glacial depositional system,
Zeit fur Geom, 76: 33-73.
Whalley, W.B., McGreevy, J.P. & Ferguson,
R.J. 1984. Rock temperature observation
and chemical weathering in the Hunza
region, Karakoram: Preliminary data. In:
Miller, K. (Ed.) . International
Karakoram Project, Cambridge University
Press, Cambridge, pp. 616-633.

Formation and Deformation of Glacial Deposits, Warren & Croot (eds) © 1994 Balkema, Rotterdam, ISBN 90 5410 096 6

Tills and non-till diamictons in glacial sequences

Joanne M. R. Fernlund
Uppsala University, Sweden

ABSTRACT: In glaciated areas diamicton sediments are often interpreted as till; it is seldom questioned whether they might be of some other origin. There are several ways in which non-till diamicton can form, including: clastic intrusions, shearing of sediments, collapse and mixing of sediments due to melting of buried ice blocks, mixing of sediments due to liquefaction, and deposition of debris-rich aquatic sediments. The examples described are associated with a glacial advance along the central part of Sweden's west coast. However with the exception of some of the sheared diamicton, most of them have never had any direct contact with the glacier. Proper identification of diamicton sediments, till or non-till diamicton, can extensively **affect** the interpretation of an area's glacial history. Critical examination of the genesis of a sediment must be made before adopting the genetic term "till" instead of the descriptive term "diamicton".

1 INTRODUCTION

Diamicton or till: in glaciated areas great portions of the diamictons are till, sediments transported and deposited by glacial ice (Dreimanis 1989). However, some sediments termed till in the past are being re-evaluated as non-till diamicton, since they do not represent direct glacial deposition. On the other hand, glacial events have been recognized which have left little evidence, either primary deposition or resulted in any substantial glacial erosion (Lagerbäck 1988a, 1988b, Lagerbäck & Robertsson 1988, Kleman & Borgström 1990). With respect to the glacial history of an area, a deformed sediment, where the pre-existing sediments have been deformed in contact with the glacier's bed, may rightly be termed till.

The purpose of this paper is to exemplify the need to critically evaluate the genetic origin of a sediment before naming it a till. Examples are presented of both non-till diamictons and of slightly deformed sediments which are interpreted as till. In general, these types of sediments are probably more common than presently recognized. Their proper identification directly effects the interpretation of the glacial history. It should

be emphasized that in practice it can be difficult to determine if a diamicton is a till or of some other origin.

2 GEOLOGICAL SETTING

The sites described are located in Halland, southwestern Sweden (Fig. 1) A recent lithostratigraphic study has revealed a need to revise the glacial history of the area. Both the traditional and the revised glacial histories are briefly described below. They differ primarily with respect to the genesis of the Halland Coastal Moraines and the time at which Halland was ice free.

2.1 *Traditional glacial history*

The deglaciation of Halland after the Weichselian Maximum occurred c. 13,500 BP in southern Halland and c. 12,600 BP in northern Halland (Lagerlund et al. 1983, Björck et al. 1988). Sea level during the deglaciation was of the order of +50 to +85 m above present, south to north respectively (Robison 1983, Påsse 1986, 1988, 1990). The Halland Coastal Moraines are des-

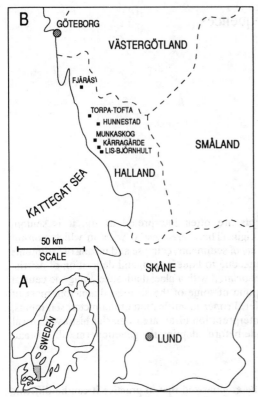

FIG. 1. Location map. The sites described in the text are marked.

cribed as end moraines deposited during the deglaciation in a subaquatic environment. They consist of a group of ridges, most of which trend sub-parallel to the present coast, and are located predominantly in central Halland, exclusively below the marine limit. They are supposed to reflect a yearly glacial retreat and a straight ice margin (De Geer 1893, Svedmark 1893, Caldenius 1942, Gillberg 1956, Hillefors 1969, 1975, 1979, Mörner 1969, Fält 1975, Påsse 1986, 1988).

2.2 Revised glacial history

The Halland Coastal Moraine Ridges are glaciotectonic in origin with some degree of diapiric intrusion (Fernlund 1988a, 1988b, 1990, 1993). The ^{14}C dates, of shells in the deformed sediments associated with the ridges, indicate that the initial deglaciation following the Weichselian Maximum occurred at c. 13,000 BP, and that the glaciotectonics occurred some time after 12,400

BP (Fernlund 1993). The glaciotectonics are interpreted to be the result of a younger glacial advance, the Halland Advance, either a surge of the ice-sheet or an expansion of a local ice cap (Fernlund 1990). Sea level at the time of the advance would have been between +40 to +15 m above present, depending on when the advance occurred (Påsse 1988, Fernlund 1990, 1993).

3 TYPES OF DIAMICTON

The types of diamicton discussed below would also be expected to occur in other glaciated areas, especially those situated below the marine limit after deglaciation. Diamicton can probably form in several other ways and in other environments than those in the examples. Some of the diamicton types described have no direct relationship to glacial activity, whereas others are closely associated with glacial activity but are not always the direct deposit of a glacier. That is to say they are not always till.

3.1 Diamictic clastic intrusions: diapirs, dikes and sills

The process of clastic intrusion could be induced by: "1. orogenic compression; 2. postorogenic isostatic movements and tension faults; 3. differences in density and thickness of sediments; and combinations of 1, 2 and 3" (Kugler 1978). This would produce diapirs containing large blocks of rocks which rose from a lower stratigraphical position. The stratigraphy of the area would be critical for development of intrusions. For example, their formation seems to require several stratigraphical layers with either an underlying aquiclude or low-density layer.

Diamictic clastic intrusions seem to be quite common in the Halland area both in association with the Halland Coastal Moraine Ridges and in glaciofluvial delta deposits. The internal composition of the intrusive bodies varies greatly, from pure clay, to a mixture of sand, gravel, and diamicton, and to pure diamicton. There are several factors which could account for the extent of compositional variation, although the most important probably are variations in the composition of both the mobilized pre-existing sediments and the sediments which were intruded.

At the Hunnestad glaciofluvial delta there are several ridges of homogeneous diamicton which

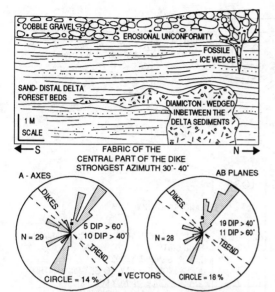

FIG. 2. A cross section from the Hunnestad Delta, western gravel pit. A dike and sill of diamicton cuts and is wedged in between the bedding. Therefore it must have intruded after the deposition of the delta. The steeply dipping clasts support an intrusive mode of formation for the diamicton.

have intruded, both as dikes and sills, into the delta sediments (Fig. 2). The diamicton in the ridges is massive and hard. The fabric analysis indicates that a large portion of the clasts dip very steeply. There is a similar intrusion of pure clay into the delta foreset beds at the Fjärås glaciofluvial delta (Fig. 3).

Sections through several of the Halland Coastal Moraine Ridges display some degree of clastic intrusion. At the Torpa-Tofta site the composition of the sediments varies between diamicton and well-sorted sediments. There is a great deal of soft sediment deformation. Intrusive sills of diamicton occur between the bedding of the sorted sediments (Fernlund 1988a fig. 5). At the Munkaskog (Fernlund 1990 fig. 2, 1993 fig. 3) and Kärragärde sites (Fig. 4) a similar mixture between diamicton and well-sorted sediments has intruded the rhythmites.

It is clear from the structural relationships at these sites (i.e. the diamicton dikes and sills cut the bedding of the sediments into which they intruded) that the intrusion occurred after the deposition of the overlying sediments. The diamicton produced during the intrusion was not in contact with the glacier and therefore is not till.

The examples of diamicton intrusions, described above, are interpreted to be associated with the Halland Advance. However, since there is no control on when the intrusions occurred, it is not possible to exclude the possibility that isostatic uplift could be responsible (Fernlund 1993).

3.2 Sheared diamicton

The production of sheared diamicton could either be due to a glacier, an iceberg or possibly sea ice. In the case of glacial formation there are two possibilities. First, the uppermost sediments would be sheared in the deforming wedge underneath the ice (Boulton 1979, Alley et al. 1986, Boulton & Hindmarsh 1987, Hart 1990). Second, shearing would occur as the result of glaciotectonics at the boundaries between fault blocks, especially imbricately thrust blocks of sediment (Fernlund 1988a, 1990). An increasing degree of shear would produce sediments with increasing degrees of deformation; from only slightly deformed to diamicton and eventually to a shearbedding (van der Wateren 1987, 1991). If the shear deformation was produced in the subglacial deforming wedge, then with respect to the glacial history, it would be justifiable to call the sediment till, whether or not it is possible to determine if it was incorporated within the glacier. Whereas, if the shear deformation was the result of mixing of sediments along fault planes, by iceberg plowing or sea ice movement, then the interpretation of till would be totally misleading to the glacial history of the area.

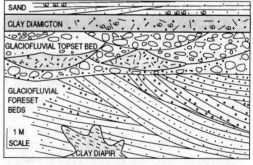

FIG. 3. A composite section of the delta at Fjärås. There are three types of diamicton represented: 1. sheared shelly, clay-rich diamicton above the topset bed; 2. liquifaction diamicton occurs as trough beds in the topset bed; 3. a massive clay diapir has intruded into the foreset beds from below.

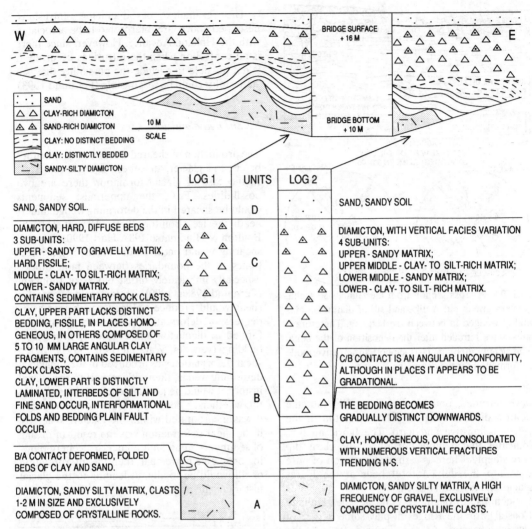

FIG. 4. Kärragärde section, trending N80°E. Two types of diamicton occur. The lower sandy diamicton (unit A) has intruded upwards into rhythmically bedded clay-sand-silt (unit B). Sheared diamicton (unit C) occurs above the clay (unit B). It varies in thickness between c. 1 m to 3 m, and matrix composition between sand and clay rich. The sheared diamicton probably formed in the deforming wedge under the glacier. The matrix variations might reflect shear bedding or variations in pre-existing bedding.

Diamicton sediments have commonly been observed along fault planes, both normal and thrust faults. At the Torpa-Tofta site, along the ice proximal side of the ridge, there is a sequence of c. 2 to 3 m thick blocks, consisting of shell bearing rhythmites which are imbricately thrust. Along one thrust zone, there is a 1 m thick clay-rich sheared diamicton which displays numerous shear planes with slickensides (Fernlund **1988a fig. 9**). Along another thrust zone, rhythmites are folded together with sandy-silty diamicton. Since

the diamicton also contains shells it is interpreted to have formed by shear mixing of the rhythmites and glaciofluvial sediments (Fernlund **1988a fig. 6**). The thrusting is interpreted to be the result of the Halland Advance. The diamicton between the blocks would never have had any contact with the glacier and therefore would not be till.

Sheared diamicton, interpreted to have been formed in a subglacial deforming wedge, occurs at several sites. At the Munkaskog site there is a c. 0.5-1 m thick clay-rich diamicton which overlies

32

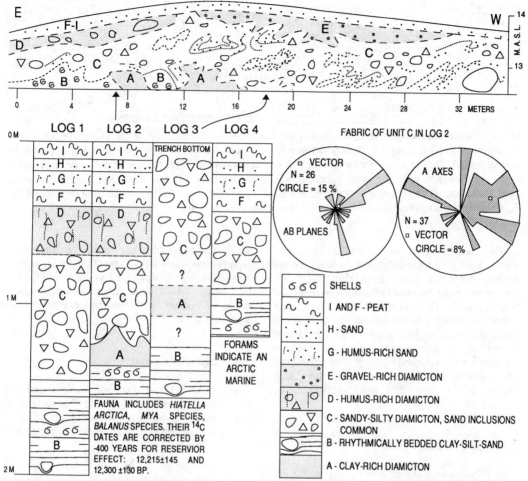

FIG. 5. A cross section roughly perpendicular through the N10°E trending ridge at Lis-Björnhult. Log 1 and 4 are located 10 m east and west of the trench, respectively; Log 3 extends from the bottom of the pit downward 4 m. There are numerous inclusions of folded laminated sand in unit C. Units A, C, D and E appear to be sediments which have been folded and sheared into diamicton in situ. The shear direction is predominantly from east towards west.

c. 5 m of rhythmites. The contact between the two is an erosional unconformity, although in places it appears to be gradational (Fernlund 1990 fig. 2). Similarly at the Kärragärde site (Fig. 4) there is c. 1.5-3 m thick diamicton with diffuse bedding, varying between sand- and clay-rich diamicton. The diamicton grades into the distinctly bedded rhythmites below and adjacent (Fig. 4, log 1 compared to log 2). At the Lis-Björnhult site (Fig. 5) the upper sediments consist of various diamicton masses (units A, C, D and E), clay rich to gravel rich, with shear towards the west.

They grade downwards into rhythmically bedded clay (unit B) (Fig. 5, log 3). At the Fjärås site (Fig. 3) there is a 1 m thick clay diamicton overlying the delta sediments. The clay diamicton is composed of small, < 1 cm, angular cubes of clay and shell fragments. At the Hunnestad Delta the sediments are only slightly deformed, in places with numerous small reverse faults and in others dilated and folded into a homogeneous sandy-gravelly sediment lacking distinct bedding (Fig. 6). If the degree of deformation had increased, then the sediments might have become diamicton.

33

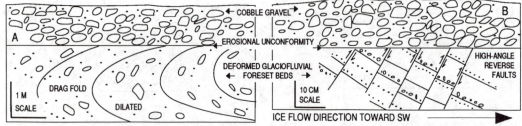

FIG. 6. Two parallel sections along the upper part of the Hunnestad Delta, eastern gravel pit. The beds below the erosional unconformity are slightly deformed: A. dilated and folded (mirrored scetch), B. high-angle reverse faulted. The deformation extends less than c. 2 m downwards. The direction of deformation is similar to the ice-flow direction. The upper cobble gravel is lithologically different than the glaciofluvial foreset beds. The erosional unconformity, which occurs in all the sections from the Hunnestad delta (Fig. 2 and Fig. 7), is interpreted to represent the younger glacial advance over the delta.

The sheared diamicton at Munkaskog, Kärra-gärde, Fjärås and Lis-Björnhult, as well as the slightly deformed sediment at Hunnestad, are interpreted to be the result of subglacial deformation due to the Halland Advance. In all of these cases the shearing is interpreted to have taken place in contact with the glacier. In some cases it is difficult to determine whether the sheared diamicton was deposited by the glacier or deformed in situ. However, with respect to the glacial history of the area, they would justifiably be termed till.

3.3 Liquefaction diamicton

The process of liquefaction mixing can form diamicton sediments. The cause of the liquefaction can include: tremors, isostatic uplift, or loading of hydrodynamically unstable sediments. Since liquefaction occurs post-depositionally in situ, the sediment would have no direct contact to a glacier and would not be till.

An example of this type of diamicton exists at Fjärås (Fig. 3). The topset bed consists of large scale trough cross beds; those which are primarily composed of cobbles are bedded, whereas those which are predominantly composed of silt and sand with very few larger clasts have been mixed into a homogeneous diamicton. This is interpreted as post-depositional mixing of the sediments within a single bed. There are distinct contacts between the troughs of diamicton and well-sorted sediment. The sheared clay diamicton overlying the topset bed suggests that a glacier advanced over the delta (Fig. 3). It is interpreted that during this advance, the hydrodynamically un-

stable silty beds within the topset beds underwent liquefaction, producing the diamicton trough beds. Although in this case the diamicton formation was initiated by a glacial advance, it was never in contact with the active glacier, and therefore is not till.

3.4 Collapsed diamicton

This process is most often the result of collapse and mixing of sediments due to the melting of interbedded, underlying blocks of ice. This is commonly recognized in glaciofluvial deltas and

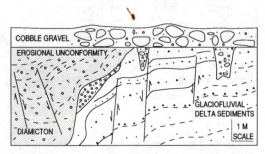

FIG. 7. A cross section along the ice-proximal side of the Hunnestad Delta, northern pit. Numerous high-angle faults suggest that the deformation is due to melting of an underlying interbedded ice block. Due to collapse, the glaciofluvial sediments have mixed to a diamicton. The deformation occurred prior to the deposition of the upper cobble gravel, which has a different lithological composition than the delta sediments. The erosional unconformity probably marks the younger glacial advance over the delta.

34

eskers where ice blocks were buried within the glaciofluvial sediments. Diamicton is formed primarily by the slumping and mixing together of the overlying glaciofluvial sediments as the ice melts. In some cases the diamicton can be traced laterally, grading into primary glaciofluvial sediments. This diamicton type would form as the buried ice melted, often a considerable time after the glacier retreated from the site. It does not represent a glacial oscillation.

An example of this is seen along the steep ice-contact slope of the Hunnestad ice marginal delta (Fig. 7). The structures suggest that a large ice block was totally buried within the delta sediments. As the ice melted the sediments slumped and mixed together forming diamicton. This diamicton is not formed in contact with a glacier and does not require a glacial advance, but instead is formed after the glacier retreated from the area.

4 DISCUSSION

There are many types of environments in which diamicton can form; in glaciated areas most of the diamicton is till, but there are several types of diamicton which are not. The above examples from Halland are often associated with a glacial advance in the area, although, with the exception of some of the shear diamicton, most of them were formed without contact to the glacier.

Diamictic clastic intrusions are probably quite common in areas with thick Quaternary sediments. They originate primarily due to density difference in the sediments, similar to salt diapirs. Isostatic uplift would cause changes in the hydrogeological conditions which could cause the hydrodynamically unstable sediments to intrude upward into overlying sediments. Other causes of clastic intrusions are earth quakes or loading of the overlying sediments. The latter could either be due to continued sedimentation producing a thick sequence of overlying sediments or possibly due to a glacial advance. These diamictic clastic intrusions are not till.

Sheared diamicton can form during faulting, such that the sediments along the fault are mixed together during movement. Another possible genesis for sheared diamicton is sea ice deformation, an example would be stamukhi sea-ice as it interacts with the underlying sediments (Reimnitz et al. 1978a, 1978b, Rearic et al. 1990). These two types of diamicton would not be till; such an interpretation would be totally misleading to the glacial history.

Some sheared diamictons are till. These typically form in direct contact with the base of the advancing glacier in the deforming wedge (Boulton 1979, Alley et al. 1986, Boulton & Hindmarsh 1987, Hart 1990).

The other two diamicton types exemplified above are liquefaction diamicton, which is formed in situ after primary deposition, and collapse diamicton, which is formed as buried ice melts. Neither is till.

There are probably several other ways in which non-till diamicton can be formed. One of the more common types, which often has caused much debate, is diamicton composed of debris-rich aquatic sediments. A typical example of the impact of a sediment's interpetation is a clay-rich diamicton in Skåne (Fig. 1). "Lund Till" (Lagerlund 1980) renamed as "Lund Diamicton" (Malmberg Persson & Lagerlund 1990) is at present both described as a glacioaquatic sediment (Houmark-Nielsen & Lagerlund 1987, Malmberg Persson 1988, Malmberg Persson & Lagerlund 1990), and as lodgement till (Holmström 1904, Ringberg 1984, 1987). Needless to say these two interpretations of "Lund Till" - "Lund Diamicton" result in two totally different glacial histories for the area, especially with respect to sea level and ice marginal position.

5 CONCLUSIONS

It is very important to be critical when interpreting a diamicton as till. Till requires the presence of a glacier at the site at the time of formation of the sediment. Although several non-till diamicton types may typically be associated with glacial events, they may not be deposited by the glacier.

Non-till diamicton is probably more common than presently assumed. Some of the geneses of these non-till diamictons are clastic intrusions, shearing of sediments, mixing during slumping of sediments due to melting of buried ice blocks, mixing of sediments due to liquefaction, and deposition of debris-rich aquatic sediments. None of these require glacial activity for their formation although they can be associated with a glacial advance. There can be radical differences in the interpretation of the glacial history of an area depending upon the interpretation of the sediments, diamicton or till.

ACKNOWLEDGEMENTS

First I want to thank the Swedish Natural Science Research Council for financing my research. I would also like to thank Olof Ohlsson and Barbro Stener, the property owners of Lis and Björnhult, for allowing the trench to be dug, and Jan Richard Larsson for digging the trench. I would like to thank all the members of the research team working on problems in Halland: Tore Påsse, Svante Björck, Anders Rapp, Rolf Nyberg, Harald Svensson, Fredrik Klingberg and Mikael Berglund, for their ideas, discussions and support. I would also like to thank all the people who have read this manuscript and given their criticism.

REFERENCES

Alley, R.B., Blankenship, B.I., Bentley, C.R. & Rooney, S.T., 1986: Deformation of till beneath ice stream B, West Antarctica. Nature 322, p. 57-59.

Björck, S., Berglund, B.E., Digerfeldt, G. & Lagerlund E., 1988: New aspects on the deglaciation chronology of South Sweden, Geographia Polonica 55, p. 37-49.

Boulton, G.S., 1979: Processes of glacier erosion on different substrata. Journal of Glaciology 23, p. 15-38.

Boulton, G.S., & Hindmarsh, R.C.A., 1987: Sediment deformation beneath glaciers: Rheology and geological consequences. Journal of Geophysical Research 92(B9), p. 9059-9082.

Caldenius, C., 1942: Gotiglaciala israndstudier och jökelbäddar i Halland. Förelöpande meddelande. Geologiska Föreningens i Stockholm Förhandlingar 64, p. 163-183.

De Geer, G., 1893: Praktiskt geologiska undersökningar inom Hallands län. Sveriges Geologiska Undersökning C-131, p. 1-38.

Dreimanis, A., 1989: Tills: Their genetic terminology and classification. In: R.P. Goldthwait and C.L. Matsch (eds.) Genetic Classification of Glacigenic Deposits. Balkema, Rotterdam, p. 17-96.

Fält, L.M., 1975: Ändmoräner, kustområdet mellan Kungsbacka och Värö. C. kursarbete, Chalmers Tekniska Högskola, Göteborgs Universitet, Geologiska Institutionen, Publication B35, p. 1-43.

Fernlund, J.M.R., 1988a: The Halland Coastal Moraines: Are they end moraines or glacio-tectonic ridges? In: D.C. Croot (ed.) Glaciotectonics. Forms and processes. Balkema, Rotterdam, p. 77-90.

Fernlund, J.M.R., 1988b: The Halland Coastal Moraines: an example of their lithostatigraphy from the ridge at Tofta, Varberg. 18. Nordiske Geologiske Vintermöde, Köpenhamn 1988, abstracts, Danmarks Geologiske Undersögelse, p. 109-110.

Fernlund, J.M.R., 1990: An introduction to the "Halland Advance" the Fennoscandian Ice Sheet or an ice cap? In: E. Lagerlund (ed.) Methods and Problems of Till-Stratigraphy - INQUA - 88 proceedings. Symposium and field trip in southern Sweden 25-30 September 1988. LUNDQUA Report 32, Department of Quaternary Geology University of Lund, p. 41-46.

Fernlund, J.M.R., 1993: The long-singular ridges of the Halland Coastal Moraines, Southwestern Sweden, Journal of Quaternary Science 8, in press.

Gillberg, G., 1956: Den glaciala utvecklingen inom sydsvenska höglandets västra randzon III. Issjöar och isavsmältning. Geologiska Föreningens i Stockholm Förhandlingar 78, p. 357-458.

Hart, J.K., 1990: Proglacial glaciotectonic deformation and the origin of the Cromer Ridge push moraine complex, North Norfolk, England. Boreas 19, p. 165-180.

Hillefors, Å., 1969: Västsveriges glaciala historia och morfologi. Naturgeografiska studier. Meddelanden från Lunds Universitets Geografiska Institution Avhandling 60, p. 1-319.

Hillefors, Å. 1975: Contribution to the knowledge of the chronology of the deglaciation of western Sweden with special reference to the Gothenburg Moraine. Svensk Geografisk Årsbok 51, p. 70-81.

Hillefors, Å. 1979: Deglaciation models from the Swedish west coast. Boreas, Vol. 8, p. 153-169.

Holmström, L, 1904: Öfversigt af den glaciala afslipningen i Sydskandinavien. Geologiska Föreningens i Stockholms Förhandlingar 26, p. 241-432.

Houmark-Nielsen, M., & Lagerlund, E., 1987: The Helsingör Diamicton. Bulletin of the Geological Society of Denmark 36, p. 237-247.

Kleman, J., and Borgström, I., 1990: The boulder field of Mt. Fulufjället, west-central Sweden - late Weichselien boulder blankets and interstadial periglacial phenomena. Geografiska Annaler 72A, p. 63-78.

Kugler, H.G., 1978: Volcanism, sedimentary, In: R.W. Fairbridge & J. Bourgeois (ed) The encyclopedia of sedimentology, Encyclopedia of earthsciences series, Vol VI. p. 854-858.

Lagerbäck, R., 1988a: Periglacial phenomena in the wooded areas of Northern Sweden - relics from the Tärendö Interstadial. Boreas 17, p. 487-499.

Lagerbäck, R., 1988b: The Veiki moraines in northern Sweden - widespread evidence of an Early Weichselian deglaciation. Boreas 17, p. 469-486.

Lagerbäck, R., & Robertsson, A-M., 1988: Kettle holes - stratigraphical archives for Weichselian geology and palaeoenvironment in northernmost Sweden. Boreas 17, p. 439-468.

Lagerlund, E., 1980: Litostratigrafisk indelning av Västskånes Pleistocen och en ny glaciationsmodell för Weichsel. LUNDQUA Report 21, Department of Quaternary Geology, University of Lund, p. 1-120.

Lagerlund, E., Knutsson, G., Åmark, M., Hebrand, M., Jönsson, L.O., Karlgren, B., Kristiansson, J., Möller, P., Robison, J.M., Sandgren, P., Ternor, T. & Waldemarsson, D., 1983: The deglaciation pattern and dynamics in south Sweden, a preliminary report. LUNDQUA Report 24, Department of Quaternary Geology, University of Lund, p. 1-7.

Malmberg Persson, K., 1988: Lithostratigraphic and sedimentological investigations around the eastern boundary of Baltic deposits in central Scania. LUNDQUA Thesis 23, Department of Quaternary Geology University of Lund, p. 1-72.

Malmberg Persson, K., & Lagerlund, E., 1990: Sedimentology and depositional environments of the Lund Diamicton, southern Sweden. Boreas 19, p. 181-199.

Mörner, N.A., 1969: The Late Quaternary History of the Kattegatt Sea and the Swedish West Coast. Sveriges Geologiska Undersökning, C 640, p. 1-487.

Påsse, T., 1986: Beskrivning till jordartskartan Kungsbacka SO. Sveriges Geologiska Undersökning, Ae 56, p. 1-106.

Påsse, T., 1988: Beskrivning till jordartskartan Varberg SO/Ullared SV. Sveriges Geologiska Undersökning, Ae 86, p. 1-98.

Påsse, T., 1990: Beskrivning till jordartskartan Varberg NO. Sveriges Geologiska Undersökning, Ae 102, p. 1-117.

Rearic, D.M., Barnes, P.W., Reimnitz, E., 1990: Bulldozing and resuspension of shallow-shelf sediment by ice keels. Implications for arctic sediment transport trajectories. Marine Geol-

ogy 91, p. 133-147.

Reimnitz, E., Toimil, L., Barnes, P., 1978a: Arctic continental shelf morphology related to sea-ice zonation, Beaufort Sea, Alaska. Marine Geology 28, p. 178-210.

Reimnitz, E., Toimil, L., Barnes, P., 1978b: Stamukhi zone processes: Implications for developing the arctic offshore area. Journal of petroleum technology, p. 982-986.

Ringberg, B., 1984: Beskrivning till jordartskartan Helsingborg SO. Sveriges Geologiska Undersökning, Ae 51, p. 1-174.

Ringberg, B., 1987: Beskrivning till jordartskartan Malmö NO. Sveriges Geologiska Undersökning, Ae 85, p. 1-147.

Robison, J.M., 1983: Glaciofluvial sedimentation: A key to the deglaciation of the Laholm area, southern Sweden. LUNDQUA Thesis 13, Department of Quaternary Geology University of Lund. p. 1-92.

Svedmark, E., 1893: Beskrifning till kartbladet Varberg. Sveriges Geologiska Undersökning, Ab 13, p. 1-82.

Wateren, van der F.M., 1987: Structural geology and sedimentology of the Dammer Berge push moraine, FRG. In: J.J.M. van der Meer (ed.) Tills and Glaciotectonics. Balkema, Rotterdam, p. 157-182.

Wateren, van der F.M., 1991: Shear zones in unlithified sediments. In: INQUA Commission on the formation and properties of glacial deposits Ireland 1991, Abstracts of papers.

Formation and Deformation of Glacial Deposits, Warren & Croot (eds) © 1994 Balkema, Rotterdam, ISBN 90 5410 096 6

The micromorphological character of the Ballycroneen Formation (Irish Sea Till): A first assessment

Jaap J.M. van der Meer & Anja L.L.M. Verbers
Fysisch Geografisch en Bodemkundig Laboratorium, University of Amsterdam, Netherlands

William P. Warren
Geological Survey of Ireland, Beggars Bush, Dublin, Ireland

Abstract

In this study a number of thin sections from diamicton facies of the Ballycroneen Formation (Irish Sea Till or related sediments) are described. The observed features are then compared to the micromorphology of basal tills, glaciomarine deposits, glaciolacustrine deposits and glaciotectonic structures. The conclusion is that the Irish Sea Till shows the greatest resemblance to basal tills, while the related sediments show features that are comparable to glaciotectonic structures. There is nothing in the thin sections that supports an origin of the Irish Sea Till as an in situ glaciomarine deposit.

Introduction

The Ballycroneen Formation (Warren, 1985) is exposed extensively along the east and south coasts of Ireland as far west as Ballybranagan in County Cork. Although the formation is a heterogenous suite of glacigenic sediments, the dominant facies exposed is a clay-rich diamicton commonly referred to as the Irish Sea Till. This diamicton is characterised by its brown, clay-silt matrix and erratics of northern and Irish Sea Basin provenance including shell fragments, shells and flint (see Warren, 1985). The Irish Sea Till is largely confined to the coastal zone, it extends at most about 12 km inland. The Irish Sea Till has traditionally been regarded as a basal till produced by glacial transportation onshore of Irish Sea Basin sediments (hence the shells; cf. Lamplugh et al, 1903). A similar interpretation has been given to other shell-bearing tills in coastal settings, not only in Ireland (Cole 1911/12; Lamplugh 1911), but also in other countries (e.g. Scotland: Gray & Brooks, 1972). Recently the Irish shell-bearing tills and associated deposits have been re-interpreted as in situ glaciomarine deposits (Eyles & McCabe 1989a, b; McCabe 1987; McCabe et al 1990). As the latter interpretation does not go unchallenged (e.g. Synge, 1981; Thomas & Summers, 1983; Harris, 1991; Warren, 1991a), but especially because of the con-

sequences of a glaciomarine origin for all aspects of glaciation (isostasy, sealevel, ice dynamics), it is important to look carefully at the evidence for the genesis of

Fig. 1. Map showing the localities where the Irish Sea Till has been sampled for thin section studies.

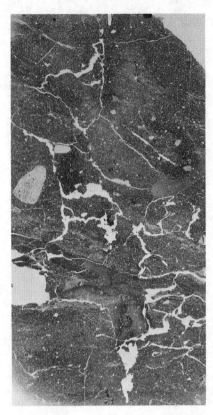

Fig. 2A. Clogga, thin section O.639 in plane light, for size see 2B.

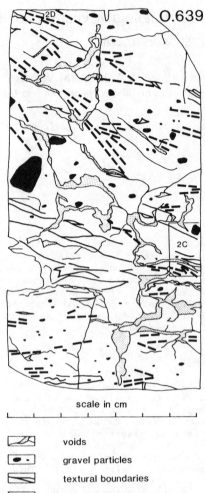

scale in cm

	voids
	gravel particles
	textural boundaries
	discrete shears

Fig. 2B. Sketch of thin section O.639 showing textural boundaries and the distribution of prominent discrete shears; rectangles indicate Figs. 2C & D.

Methods and terminology

the Irish Sea Till.
The evidence for the origin of the Irish Sea Till is in the formation itself. It should first of all be studied in detail in the field to check whether the macroscopic sedimentary and structural features are compatible with either explanation. Such field studies should be supported by laboratory studies, like e.g. grain size analyses, determination of age and species of the shell fragments, or the analyses of microfossils.

A technique that has only recently been introduced in the study of glacial sediments is thin section analyses or micromorphology (van der Meer, 1987, 1992; van der Meer & Laban, 1990; van der Meer et al, 1983, 1986, 1992; Rappol et al, 1989). In this paper we will present preliminary results of the inspection of a small number of thin sections of Irish Sea Till and compare these with what is known of the micromorphology of tills and other glacigenic deposits.

After studying and recording a wide variety of sections (e.g. Verbers, 1989), representative sites were selected for sampling. Most of the samples were taken in metal containers following the method described in van der Meer, 1992. After transport to the laboratory in Amsterdam, samples were air-dried, impregnated with an unsaturated polyester resin (full impregnation in a vacuum chamber), cut, mounted on glass supports and polished to ca. 20 μm thickness (Murphy, 1986; van der Meer, 1987, 1992). The samples were studied un-

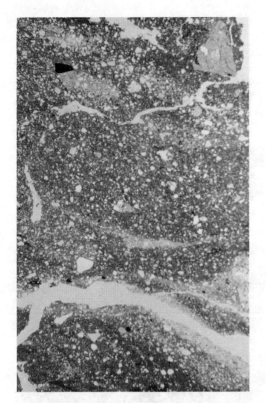

Fig. 2C. Detail of O.639 in plane light, showing textural differences; field of view is 18.0 mm.

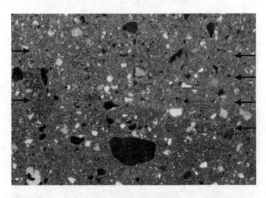

Fig. 2D. Detail of O.639 under cross-polarised light, showing discrete shears (arrows); field of view is 5.6 mm.

der an ordinary petrographic microscope with a low magnification.
The terminology used in the description of the thin sections is the terminology that has been developed by pedologists (Brewer,

1976; see also van der Meer 1987, 1992). The terminology with regard to the origin of the Formation in question will follow Dreimanis (1989) for till. The fact that a till contains shells or is purely derived from marine sediments does not make it a glaciomarine deposit. The description of the glaciomarine environment by Dowdeswell & Scourse (1990) and their statement that glaciomarine environments include 'sediment .. deposited in the sea after release from glacier ice' seems to leave little to question. After all 'deposited in the sea' must be taken literally. However, the addition that this includes grounded tidewater fronts, may lead to wrong conclusions, because it may be read as implying that material deposited directly from the ice underneath a <u>grounded</u> tidewater front is a glaciomarine deposit. By definition (Dreimanis, 1989) this is till, whereas the material deposited in front of such a glacier (in the sea) is glaciomarine. The complicating fact that – during ice retreat – the ice of the grounded tidewater glacier which deposited the till is replaced by seawater, does not make the till a glaciomarine deposit.

Description of sites and samples

Clogga, general

Sample (O.639) was collected at Clogga, South of Arklow on the Irish east coast (Fig. 1). The sample was taken from the central part of the shelly till sequence where it overlies the marine platform, about 300 m S of the stream that meets the coast immediately north of Clogga Head. The locality was visited during the 1991 INQUA field trip (Warren, 1991b). In this locality the lower 2 m of till is clast-rich, while the main till body is in general clast deficient (Huddart, 1981). In the original section recorded by Synge (Warren, 1991b) laminated silts and fine sands are indicated just to the N of the sample site. Such horizontal beds (up to 20 cm thick) were also present at the sampling site. Grain-size analyses of the till at the sample site gave a sa(nd)-si(lt)-cl(ay) composition of 46.5-28.5-25.0, a carbonate content of 11.2 %.

Clogga, sample description

At first sight this thin section looks like a very uniform brown, clayey deposit. It shows a fairly large number of cracks without apparent pattern. They are most likely to be the result of drying (Fig. 2A). There are few large pebbles, and coarse sand seems to be largely absent. Inspection under the microscope shows that there are distinct differences in grain-size (Fig. 2C). The major part of the thin section is indeed clayey and uniform, but we find intercalated laminae and lenses of silt and fine sand. When the boundaries (which are usually clear) between different grain-sizes are traced, a subhorizontal pattern emerges (Fig. 2B). The lower half of the thin section shows more of these boundaries than the upper half. X-rays of the remaining part of the sample (van der Meer et al, 1983) also show a faint subhorizontal lamination.

Another important element of micromorphology is the (re-) orientation of fines (mainly clay, often in packages or domains) which is called the plasmic fabric. In this sample we find a clear difference in the development of the plasmic fabric in the two halves. The upper half (with fewer textural boundaries) shows a clear bimasepic plasmic fabric: the oriented domains form distinct lines in two directions (Fig. 2B). These lines are due to re-orientation of the clays, (resulting in a high birefringence and hence visibility under cross-polarised light) and must be regarded as discrete shears (Fig. 2D). In the lower half these discrete shears are less obvious. The latter does not necessarily indicate absence of shear, it may reflect masking due to a high amount of finegrained carbonates, because these distort the polarising effect of the microscope.

When we plot the most obvious discrete shears (Fig. 2B) we can see that both directions are more or less parallel to the orientations of the textural boundaries. In the lower part of the sample the orientation is slightly different from that in the upper part, demonstrating the (shear) relation between the two.

Ballycroneen and Ballycotton, general

A group of 6 samples was collected on the south coast (Fig. 1), 3 each at Ballycroneen and at Ballycotton (Verbers, 1989).

At both sites a similar stratigraphy has been recorded: local till forms the top of the sequence and overlies the Irish Sea Till. Usually there is a sharp contact between the two, although sometimes a loamy windblown deposit is intercalated. In some places the compact, shelly Irish Sea Till overlies head (gelifucted diamicton), also with a sharp contact.

At Ballycroneen the first sample (O.808) was taken across the transition between Irish Sea Till and overlying (windblown) loam. The second (O.809) was taken from

Fig. 3A. Ballycroneen, thin section O.808 in plane light, scale bar is 1 cm.

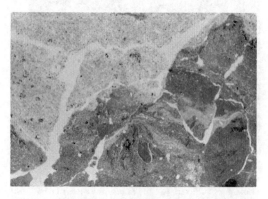

Fig. 3B. Detail of O.808 in plane light, showing the boundary between the loam and the till; field of view 18.0 mm.

Fig. 3C. The same detail seen under cross-polarised light, notice plasmic fabric.

the Irish Sea Till itself, while the last (Mi.152) was taken from a wedge of Irish Sea Till penetrating the underlying head. Although the silt content of the matrix is fairly consistent both the clay and sand

Fig. 4A. Ballycroneen, thin section Mi.152 in plane light, scale bar is 1 cm.

Fig. 4B. Detail of textural differences in sample Mi.152 in plane light; field of view is 18.0 mm.

Fig. 4C. Detail of Fig. 4B, showing the strong plasmic fabric and parallel orientation; cross-polarised light; field of view is 9.0 mm.

element vary considerably (sa-si-cl: 27.6/ 41.0 - 43.3/47.2 - 15.8/25.2; 3 samples) at Ballycroneen. At Ballycotton the matrix texture is more uniform (sa-si-cl: 26.8/ 28.5 - 48.8/53.3 - 19.1/22.7; 3 samples). The carbonate content varies between 2.8 and 14.4 % at Ballycroneen and between 5.8 and 16.3 % at Ballycotton; the lower values may well be due to leaching or reworking.

At Ballycotton the first sample was collected from an about 10 cm thick stratified bed between the head and the Irish Sea Till. The other two samples have been taken from a part of the till where it demonstrates conjugate sets of faults. At this locality the Irish Sea Till is overlain by a diamicton containing elements associated with both the Irish Sea and local (Garryvoe) tills.

Ballycroneen, sample description

Sample O.808, which was taken across the boundary between the eolian loam and the Irish Sea Till, clearly shows this lithological partition (Fig. 3). Apart from some minor inclusions of clayey material, the upper loamy part is homogeneous. These irregular clayey lenses have been dragged into the loam. Mixing may have taken place when the site was overridden by ice depositing the local (Garryvoe) till. Small scale faulting in the top of the Irish Sea Till (Figs. 3B,C) seems to point to such an effect. The Irish Sea Till itself is not homogeneous: there is a large number of silt and clay lenses of an irregular shape (Fig. 3B) as well as clay pebbles. There is a strong tectonic imprint on the Irish Sea Till itself. This is evidenced by strongly developed discrete shears, which, in some units, can be described as a trimasepic plasmic fabric (orientations in three directions) although it usually shows a masepic plasmic fabric. A lattisepic (two more or less perpendicular directions) plasmic fabric has also been noticed. Most clayey parts seem to be in some stage of brecciation. As the site has clearly been influenced by later overriding, the plasmic fabric in the Irish Sea Till here may also have been caused by the overriding and need not be a primary feature of the till.

This can be tested in the thin sections which have been collected at lower levels of the Irish Sea Till. Sample O.809 was taken from a clayey part of the till. It is of uniform grain-size and shows only a weakly developed skelsepic plasmic fabric. However, there is a distinct orientation of small (silt) particles, parallel to the

Fig. 5. Ballycotton, thin section O.806 in plane light; scale bar is 1 cm. Notice the prominent low-angle shear structure (lower right to upper left, arrows) and the boudinaging of fine grained beds.

surface of large grains. Both point to rotational movement of the particles.

The last sample (Mi.152; Fig. 4A) at this site was taken from a till wedge and can thus be expected to show a strong fabric. Although this is not always the case (van der Meer, 1982), sample Mi.152 does indeed show a strong plasmic fabric in the more clayey parts as well as distinct discrete shears (Figs. 4B,C). This latter sample clearly demonstrates that the Irish Sea Till at this site was forcefully injected into the underlying head. The wedging of till cannot be related to the later over-riding of the site.

Taking all these observations together we tend to the conclusion that the Irish Sea Till at Ballycroneen has been deposited as a basal till and cannot be considered a reworked/reoriented (much less an in situ) glaciomarine deposit.

Ballycotton, sample description

The first sample (O.806; Fig. 5) at this site was taken from a stratified bed between Irish Sea Till and underlying head. Any deformation is most likely related to the emplacement of the Irish Sea Till. The top of the sample consists of a loamy material which has been strongly sheared as evidenced by prominent structures like imbricated clay pebbles. The main part of the sample consists of alternating sub-horizontal bands of variable grain size; minute shell fragments are also present. Sometimes elongated grains show a band-parallel orientation. Towards the base of the sample till layers are present, which sometimes show boudinaging. At large magnifications one can see a weakly developed unistrial plasmic fabric. Altogether the observations in sample O.806 seem to indicate a (waterlain) sedimentary sequence which was sheared when the overlying Irish

Sea Till was deposited.

Samples Mi.153 and Mi.154 were collected from the Irish Sea Till, a little bit further North where it shows conjugate sets of faults. The two samples show some differences in grain-size in the sense that sample Mi.154 is less homogeneous than Mi.153. Also in the development of the plasmic fabric the samples show differences. Sample Mi.153 shows a weakly developed plasmic fabric, against a stronger development in Mi.154. Similar to what we have seen at Ballycroneen, sample Mi.153 does show a distinct parallel orientation of fine (silt) grains to the surface of larger grains.

Fassaroe, general

Two samples (Mi.595 and Mi.596) were collected at Fassaroe near Enniskerry, 2.5 km inland from the east coast and about 18 km south of Dublin (Fig. 1). Both samples come from the opposing sides of the core of a slightly overturned fold in Irish Sea Till. The core contains silty, laminated deposits with a large number of clay pebbles. However uniform silty and clayey bands occur as well. This whole unit is strongly deformed. The grain size of the silty material is (sa-si-cl): 21.6 - 68.9 - 9.5, with a carbonate content of 16.5 %.

Fassaroe, sample description

The two samples collected at this site are very similar. This was to be expected since both come from the same fold core. The thin sections show that the material is very heterogeneous and consists of a great variety of thin beds (Fig. 6A). These may consist of sand or silt in which grading may be present; diamictic beds do not really occur. Throughout the samples many soft sediment pebbles can be observed.

Fig. 6A. Fassaroe, thin section Mi.596 in plane light, scale bar is 1 cm.

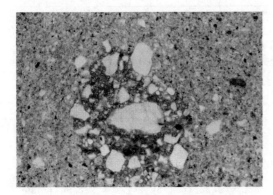

Fig. 6B. Fassaroe, detail of thin section Mi.595 in plane light, showing till pebble with a circular structure of skeleton grains; field of view is 3.5 mm.

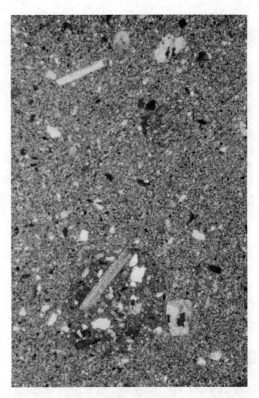

Fig. 6C. Fassaroe, detail of thin section Mi.595 in plane light, showing small till pebble containing an elongate shell fragment; field of view 4.5 mm.

Most of the pebbles consist of clay, some consist of silt, while a smaller number consists of diamicton (till pebbles, see van der Meer 1987, 1992). At least one till pebble contains a shell fragment (I-rish Sea Till, Fig. 6C). The clay pebbles demonstrate a strong plasmic fabric, which can be described as omnisepic or masepic. The till pebbles show only a weakly developed skelsepic plasmic fabric. However, some of the till pebbles instead show a circular arrangement of skeleton grains (Fig. 6B).

Structurally the samples show a large number of faults (Figs. 6A,D). On the microscale these are regularly associated with water-escape structures. The faults do not show up by a high birefringence. This is

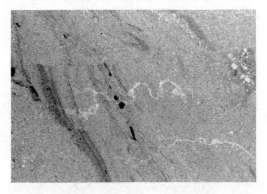

Fig. 6D. Fassaroe, detail of thin section Mi.596 in plane light, showing part of one of the faults, notice the clay pebble that has been cut off; field of view is 18.0 mm.

Fig. 7A. Shanganagh, thin section Mi.599 in plane light, scale bar is 1 cm.

Fig. 7B. Detail of thin section Mi.599 in plane light, showing drag structure outlined by clay 'lines'; field of view is 18.0 mm.

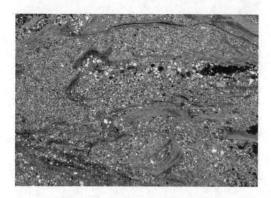

Fig. 7C. Same view as Fig. 8B, cross-polarised light, notice strong plasmic fabric in clay 'lines'

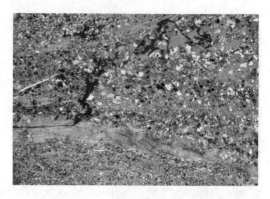

Fig. 7D. Detail of thin section Mi.599 under cross-polarised light, showing brecciation of clay 'line'; notice the presence of needle-shaped shell fragments; field of view is 18.0 mm.

most likely due to a lack of clay in many beds. Alternatively it may be due to a high content of fine-grained carbonates, because the latter strongly interfere with the polarisation of the light.

Shanganagh, general

The final sample (Mi.599) referred to in this study was taken across a shear plane overlying the basal till unit of the coastal cliff at Shanganagh (Fig. 1), in the southern suburbs of Dublin (about 1550 m north of the point at which Corbawn Lane reaches the coast). The sampling site, recently described by Eyles & McCabe (1989-a), abounds with deformational structures (Warren, 1991b). Since Eyles & McCabe do not mention these structures it seemed worthwhile to sample one of the shear planes in order to study it in detail. The grain size of this material has not yet been studied.

Shanganagh, sample description

The sample, which is situated vertically across a shear plane, consists of an alternation of sandy beds of different grain size. These sandy beds are separated by thin and discontinuous clay 'lines' The latter display a complicated, splaying pattern (Fig. 7A).
The clay lines show a very strong plasmic fabric (Figs. 7B,C), which is best described as omnisepic. Many clay lines show a kinking plasmic fabric, while several lines show intense brecciation (Fig. 7D). All the microscopic evidence points to the occurrence of intense shearing. Macroscopically this is also demonstrated by drag

structures (Figs. 7B,C). The sample contains a large number of (in cross section) needle-shaped shell fragments (Fig. 7D). In general these line up very well with the overall structure.

Discussion

When reviewing the descriptions of the thin sections, it is clear that all of the samples show distinct evidence of shear. In some this is by discrete shears, in others by different expressions of rotational movement (skelsepic or lattisepic plasmic fabric; parallel orientation of fine grains to surface of large grains; circular structure). In all cases it has to be decided whether this evidence of shear is related to a glacial stress field or whether it may have been caused by some other agent (e.g. slump).
When we want to assess the features described from the thin sections of Irish Sea Till, we can compare them to the features described from other glacigenic environments. As such we can mention 1) basal tills 2) glaciomarine deposits 3) glaciolacustrine deposits, and 4) glaciotectonic structures.

1) **Basal tills.** The structures which are common in basal tills have been described in a number of papers (van der Meer, 1987, 1990, 1991/92; van der Meer & Laban, 1990; van der Meer et al, 1983, 1986) and can be summarised as follows. Depending on the clay content (and for visibility also the carbonate content) basal tills show all or some of the following: deformed (e.g. stretched, broken) and sheared soft sediment clasts; discrete shears; skelsepic or lattisepic plasmic fabrics; orientation of fine (silt) particles parallel to the surface of large grains; a circular arrangement of skeleton grains, or crushed quartz grains. These structures can be found throughout the till deposit, though not necessarily all at the same spot (van Ginkel, 1991).

2) **Glaciomarine deposits.** This group of deposits has been relatively little studied. Samples have been obtained from different depths and environments in the Barents Sea (S of Nordaustlandet), but these samples have not yet been published (van der Meer & Solheim, in prep). However, from preliminary observations it is clear that thin sections from the glaciomarine environment show very few if any of the features mentioned for basal tills. When shearing is observed, it is confined to relatively thin zones and most likely this is related to iceberg ploughing. Similar observations have been made in thin sections from the Lund Diamicton, a sediment body on the southwest coast of Sweden and at least as enigmatic as the Irish Sea Till (Lagerlund & van der Meer, 1990; in prep). The samples from the Lund Diamicton show that it is a very inhomogeneous deposit and that the soft sediment inclusions have not been tectonised. They furthermore show only weakly developed plasmic fabrics and hence there is very little evidence of shear.
A number of samples from the Barents Sea show abundant microfossils, but without further study it is difficult to suggest which beds or environments can be typified by their occurrence.

3) **Glaciolacustrine deposits.** This group has not been studied as extensively as have basal tills. However some general observations have been published (van der Meer, 1985), while more specific data on glaciolacustrine (both deformed and undeformed) deposits in Patagonia can be found in van der Meer et al (1990a, b; 1992). The findings in these reports are supported by unpublished data. In general it can be stated that the micromorphology of glaciolacustrine sediments is very similar to that of glaciomarine sediments. It shows few of the features described for basal tills, and in ice-dammed lakes shearing may well have been produced by iceberg ploughing. The main difference between these and glaciomarine deposits is that clay laminae often show a slightly higher and original birefringence. This higher birefringence must be due to the different orientation of the clay platelets as a result of differences in settling in salt and fresh water. Another difference between glaciomarine and glaciolacustrine deposits is related to the fact that in many (small) lakes the work of currents is usually lacking or only weakly developed. As a result the chance of finding graded bedding in glaciolacustrine deposits is quite high. This is also true on a microscale.

4) **Glaciotectonic structures.**
The few examples of glaciotectonic structures that have been studied and published (van der Meer, 1987; van der Meer et al, 1985; 1992), supported by unpublished observations demonstrate that deformation by tectonism in most cases leads to strongly developed discrete shears around, and to boudinaging of, thin beds in the actual shear plane/zone. These structures resemble some of the better developed features described for basal tills. This is not

surprising since many of the structures in basal tills are produced by bed deformation, which is a tectonic action.

When we compare the micromorphology of the Irish Sea Till to what is stated for the micromorphology of different glacigenic sediments it is clear that the Irish Sea Till shows the greatest resemblance to basal tills. The structures which have been described for the Irish Sea Till samples, i.e. skelsepic plasmic fabric (in different strengths) throughout each one, discrete shears in more clayey parts (= tectonised inhomogeneities); fine grains parallel to the surface of coarse grains, all support this view. When compared to the glaciomarine deposits the evidence for shearing in the Irish Sea Till is stronger than for the Barents Sea samples. On the other hand, the inhomogeneity which is typical for the Lund Diamicton, is absent in the Irish Sea Till. Microfossils were observed only once or twice in the thin sections of Irish Sea Till. Given that macrofossils of Early Pleistocene and Tertiary age are common in sediments associated with the Irish sea Till there is no reason to assume that the micro-fossils are in situ.

The micromorphology of our thin sections do not provide any indication that these diamictons might be of glaciolacustrine origin.

The observation that the Irish Sea Till shows widespread evidence of shearing is furthermore supported by the micromorphology of samples from deposits and/or structures associated with it. Tne wind-blown loams from Ballycroneen and Ballycotton, the laminated fold core at Fassaroe, all show discrete shears and brecciation of clayey beds. The structures described in these samples are the same as those observed in the samples from glaciotectonic sites that we studied elsewhere (van der Meer, 1987; van der Meer et al, 1985; 1992).

It can thus be concluded that this preliminary study of thin sections of Irish Sea Till supports an origin as a basal till as opposed to an origin as an in situ glaciomarine deposit. The Irish Sea Till shows many changes in facies and in that sense this study has not been exhaustive. We will continue the study of the Irish Sea Till, especially in comparison to sediments of similar or supposedly similar origin.

Acknowledgements

The authors would like to thank Cees Zee-gers for the preparation of the thin sections and to Chris Snabilié, Mark Ydo and Frans Bakker for preparing the illustrations.

References

Brewer, R. 1976 Fabric and mineral analysis of soils. R.T. Krieger, Huntington.

Cole, G.A.J. 1911/12 Glacial features in Spitsbergen in relation to Irish geology. Proc. R. Irish Acad. 29: 191-208.

Dowdeswell, J.A. & J.D. Scourse 1990 On the description and modelling of glacimarine sediments and sedimentation. In J.D. Dowdeswell & J.D. Scourse (eds). Glacimarine environments: processes and sediments. Geol. Soc. Sp. Publ. 53, 1-13.

Dreimanis, A. 1989 Tills: their genetic terminology and classification. In R.P. Goldthwait & C.L. Matsch (eds). Genetic classification of glacigenic deposits, p. 17-83. Rotterdam: Balkema.

Eyles, N. & A.M. McCabe 1989a Glaciomarine facies within subglacial tunnel valleys: the sedimentary record of glacio-isostatic downwarping in the Irish Sea Basin. Sedimentology 36: 431-448.

Eyles, N. & A.M. McCabe 1989b The Late Devensian (< 22,000 BP) Irish Sea Basin: the sedimentary record of a collapsed ice sheet margin. Quat. Sci. Rev. 8: 307-351.

van Ginkel, M. 1991 Vertikale variabiliteit in de micromorfologie van dikke till afzettingen in Nederland. Doctoral Dissertation, University of Amsterdam (unpublished).

Gray, J.M. & C.L. Brooks 1972 The Loch Lomond Readvance moraines of Mull and Menteith. Scott. J. Geol. 8: 95-103.

Harris, C. 1991 Glacial deposits at Wylfa Head, Anglesey, North Wales: evidence for Late Devensian deposition in a non-marine environment. J. Quat. Sci. 6: 67-77.

Huddart, D. 1981 Pleistocene foraminifera from south-east Ireland - some problems of interpretation. Quat. Newsl. 33: 28-41.

Lagerlund, E. & J.J.M. van der Meer 1990 Micromorphological observations on the Lund Diamicton. In E. Lagerlund (ed). Methods and problems of till stratigraphy - INQUA 88 Proc. LUNDQUA Rep. 32: 37-38.

Lamplugh, G.W., J.R. Kihoe, A. McHenry, H.J. Seymour & W.B. Wright 1903 The geology of the country around Dublin (Explanation of sheet 112). Mem. Geol. Survey Ireland.

Lamplugh, G.W. 1911 On the shelley moraine of the Sefströmglacier and other Spitsbergen phenomena illustrative of British glacial conditions. Proc. Yorksh. Geol. Soc. 17: 216-241.

McCabe, A.M. 1987 Quaternary deposits and glacial stratigraphy in Ireland. Quat. Sc. Rev. 6: 259-299.

McCabe, A.M., N. Eyles, J.R. Haynes & D.Q. Bowen 1990 Biofacies and sediments in an emergent Late Pleistocene glaciomarine sequence, Skerries, east central Ireland. Marine Geology 94: 23-36.

van der Meer, J.J.M. 1987 Micromorphology of glacial sediments as a tool in distinguishing genetic varieties of till. Geol. Surv. Finland Spec. Pap. 3: 77-89.

van der Meer, J.J.M. 1990 Microscopic evidence of subglacial deformation. In R. Aario (ed.). Abstracts IIIrd International Drumlin Symposium, Oulu. Res Terrae A3: 23.

van der Meer, J.J.M. 1992 Micromorphology. In J. Menzies (ed.). Glacial environments - processes, sediments and landforms, Pergamon, in press.

van der Meer, J.J.M. & C. Laban 1990 Micromorphology of some North Sea till samples, a pilot study. J. Quat. Sci. 5: 95-101.

van der Meer, J.J.M., J.O. Rabassa & E.B. Evenson 1990a Sedimentology and micromorphology of glacigenic deposits in northern Patagonia, Argentina. In E. Lagerlund (ed.). Methods and problems of till stratigraphy - INQUA 88 Proc. LUNDQUA Rep. 32: 6-8.

van der Meer, J.J.M., J.O. Rabassa & E.B. Evenson 1990b Estudios micromofologicos de depositos glacigenicos en Patagonia septentrional. III Argentine Sedimentological Symposium, Abstracts, San Juan, 352-357.

van der Meer, J.J.M., J.O. Rabassa & E.B. Evenson 1992 Micromorphological aspects of glaciolacustrine sediments in northern Patagonia, Argentina. J. Quat. Sci. 7:

van der Meer, J.J.M., M. Rappol & J.N. Semeyn 1983 Micromorphological and preliminary X-ray observations on a basal till from Lunteren, The Netherlands. Acta geol. Hisp. 18: 199-205.

van der Meer, J.J.M., M. Rappol & J.N. Semeyn 1986 Sedimentology and genesis of glacial deposits in the Goudsberg, Central Netherlands. Meded. Rijks Geol. Dienst 39-2: 1-29.

Murphy, C.P. 1986 Thin section preparation of soils and sediments. AB Academic Publ. Berkhamsted.

Rappol, M., S. Haldorsen, P. Jørgensen, J.J.M. van der Meer & H.M.P. Stoltenberg 1989 Composition and origin of petrographically-stratified thick till in the northern Netherlands and a Saalian glaciation model for the North Sea basin. Meded. Werkgr. Tert. en Kwart. Geol. 26: 31-64.

Synge, F.M. 1981 Quaternary glaciation and changes of sea level in the south of Ireland. Geol. & Mijnbouw 60: 305-315.

Thomas, G.S.P. & A.J. Summers 1983 The Quaternary stratigraphy between Blackwater Harbour and Tinnaberna, County Wexford. J. Earth Sci. R. Dublin Soc. 5: 121-134.

Verbers, A.L.L.M. 1989 The glacial geology of southeast Co. Cork, Ireland. Doctoral Dissertation, University of Amsterdam (unpublished).

Warren, W.P. 1985 Stratigraphy. In K.J. Edwards & W.P. Warren (eds). The Quaternary history of Ireland, p. 39-65. London: Academic Press.

Warren, W.P. 1991a Fenitian (Midlandian) glacial deposits and glaciation in Ireland and the adjacent offshore regions. In J. Ehlers, P. Gibbard & J. Rose (eds). Glacial deposits in Great Britain and Ireland, p. 79-88. Rotterdam: Balkema.

Warren, W.P. (comp) 1991b Ireland 1991: Field guide for excursion. Geological Survey of Ireland, Dublin.

Formation of sorted glacigenic sediments

Formation and Deformation of Glacial Deposits, Warren & Croot (eds) © 1994 Balkema, Rotterdam, ISBN 90 5410 096 6

Controls on sedimentation in the Elsterian proglacial lake, Kleszczów Graben, central Poland

Dariusz Krzyszkowski
Geographical Institute, University of Wrocław, Poland

ABSTRACT: The Early Pleistocene (Upper Elsterian) glaciolacustrine suite of the Kleszczow Graben, central Poland is several times thicker than suites of the same age beyond the trough. They were deposited in the deep, tectonically-created basin, in part meromictic. The sediments represent typical sequence for deep, well thermally stratified lakes, but the bottom facies are interbedded with the gravity flow sediments, mostly large bodies of diamictons. These were formed on the active fault scarps within the lake basin. Simultaneously, areas lying beyond the trough are characterized by shallow lake/deltaic sedimentation in a large, ice-contact lake, with no tectonic influence.

1 INTRODUCTION

Glacial lakes are simply classified into two types: ice-contact lakes and non-contact distal lakes supplied indirectly by proglacial braided stream (Eyles & Miall, 1984; Ashley et al., 1985). Such lakes differ in physical characteristics but they each indicate strong lateral zonality of the sedimentary processes, that depents on distance from the glacier and/or delta front (Ashley, 1989). Tectonic stability is usually assumed during sedimentation. Moreover, a majority of studies have been concentrated on lakes formed during deglaciation rather than during glacial advance.

In central Poland, Pleistocene glacial lakes were formed most often during the advances of the ice-sheets. This is due to the topography which is inclined generally to the north, i.e. in the direction of advancing ice. Numerous lakes of different sizes formed between the transgressing glaciers and uplands to the south (Rozycki, 1967). They are supplied both by sub- and proglacial streams, as well as by nonglacial rivers flowing from the south. Another unusual circumstance is, that central Poland is located mostly in the tectonically unstable area of the Central Polish Placanticlinorium, a part of the young, Palaeozoic-Mesozoic Platform. This region contains deep and narrow, tectonic grabens with high rates of subsidence, whereas adjacent areas are characterized by a relatively less subsidence or even by uplift. Clearly these

factors must have influenced sedimentary processes within the proglacial lakes. First of all, the transgressive lakes are expected to be in great part the ice-contact lakes. Also, their life span is usually shorter in comparison to the deglaciation lakes of comparable size because their existence depends on the presence of ice. Finally, sedimentation within lakes is highly influenced by the changing subsidence rates. In an extreme case, this can completely disturb the typical proximal-distal character of the lake sedimentation.

Good examples of the tectonically influenced proglacial lake sedimentation are found within the Kleszczow Graben near Belchatow (Fig. 1). The graben has parallel border faults and a more than 200 m deep trough that was active from the end of Palaeogene up to Middle Pleistocene. The trough deposit are exposed in part of the Belchatow lignite quarry (200 m deep and area about 20 km). They represent a sequence of Miocene lignite, Miocene and Pliocene fluvial and lacustrine deposits and Pleistocene glacial and interglacial or interstadial fluviolacustrine suites (Krzyszkowski, 1991)(Fig. 1). Additionally the sediments within the Kleszczow Graben can be sub-divided into two structural units: the lower, tectonically deformed Belchatovian unit and the upper, non-deformed unit (Krzyszkowski, 1989a). The first comprises Tertiary and the lowermost part of Pleistocene sediments. The last one comprises the uppermost Pleist-

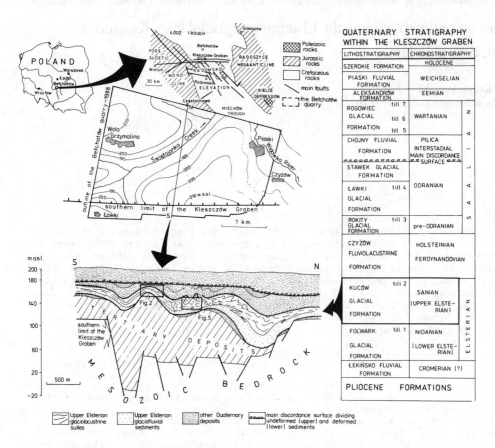

Fig. 1. Location of the Kleszczow Graben and the Belchatow quarry.
Stratigraphy of the trough sediments.

ocene sediments which lie sub-horizontally
above the main, angular discordance surface
(Fig. 1).

Four glacial formations: Folwark, Kucow,
Lawki and Stawek, which belong to the tectonically deformed structural unit, comprise
thick glaciolacustrine suites. Their thickness is five to ten times greater than the
lake sediments of the same age beyond the
graben (Baraniecka & Sarnacka, 1971). Moreover, they contain a distinctive facies association and distribution within the lake
basin. The Upper Elsterian Kucow Formation
is especially thick reaching on average 60
m and up to 100 m of glacial sediments, including 20-50 m thick glaciolacustrine unit
(Fig. 1).

The goals of this study is describe the
lithofacies association developed during
ice advance in an ice-contact glacial lake
and asses the effects of rifting tectonics
on sedimentary processes and deposits, with
examples from the Kucow Formation of the
Kleszczow Graben.

2 STRATIGRAPHY

The Kucow Formation in the Kleszczow Graben
contains four members which represent a glacial transgressive sequence. These are,
from bottom to top:
- a lower glaciofluvial sands and gravels:
comprised of cross bedded pebble sands,
coarse and medium sands. Trough cross
bedding prevails, although the planar
sets are more frequent in the upper part
of the member. In sands, there are
observed 0.5-1.0 m thick and laterally
discontinuous (from 1-2 m to several metres) gravel beds. Gravels are strongly
imbricated and they are often overlain by
thin layers of mud and/or rippled sand.
The lowermost member represents a glacial
outwash sediment that was deposited most
probably by streams of the Donjeck-type,
with well developed longitudinal bars
(Miall, 1977, 1978, 1984). This transfor-
med into the Platte-type braided stream
with deposition mostly on the transverse

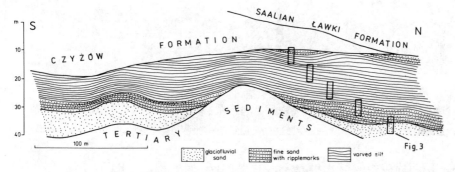

Fig. 2. The sediment succession in the marginal zone of the Elsterian proglacial lake of the Kleszczow Graben. Location of the section is in Fig. 1.

bars in the upper part of the member (Smith, 1971, 1974; Miall, 1977, 1978, 1984).
- a middle glaciolacustrine suite: a both vertically and laterally changeable sedimentary suite. Generally it can be sub-divided into central and marginal zones. The first occupies the narrow zone in the central part of the graben and is comprised of mostly varved clays and other rhythmites as well as massive sands, massive silts and diamictons. The marginal zone, which occupies the rest of the graben, is comprised of planar cross bedded sands, horizontally bedded sands, thick sets of fine sands with climbing ripple cross lamination and silty-clayey silty rhythmites ("varved silts").
- an upper glaciofluvial sands: is comprised mostly of trough and planar cross bedded pebble sands and coarse sands. The gravel beds of the large longitudinal bars have not been observed. Most probably the member represents an outwash sequence of the mixed Donjeck/Platte-type stream.
- an uppermost glacial till: is a massive till with frequent erratics and lenses of sorted sediments.

The thickness of the glaciolacustrine suite reflects simply the changeable subsidence rates within the graben, reaching 30-50 m in its central and northern parts and only 20-30 m in the southern part (Fig. 1). In areas beyond the graben the suite under discussion is always less than 15 m thick. Moreover, the glaciolacustrine sediments are deformed. Regular anticlines and synclines with wavelengths from 300-800 m and amplitude up to 50 m occur in the northern and southern zones of the trough. The central zone of the Belchatow quarry is an extensive synclinorium about 0.5-1.0 km in width. A variety of minor folds are superimposed on the major structure, with inclined, overturned and recumbent folds, as

well as diapirs and shear zones. No real pattern of deformation has been recognized. The central synclinorium occurs on the axis of the central elevation in the graben (Fig. 1), and this is coincident with the pattern of deformation has been recognized. The central synclinorium occurs on the axis of the central elevation in the graben (Fig. 1), and this is coincident with the central zone of the lake basin within the Belchatow quarry.

3 DESCRIPTION AND INTERPRETATION OF THE LAKE SEDIMENTS

3.1 Marginal zone

The sediments have been described in detail in the southern part of the Kleszczow Graben (Fig. 1). These form a 30 m thick silty-sandy sequence which indicates strong vertical lithological variability, but this is laterally homogenous over large distances (Fig. 2). Three main units can be recognized, namely the lower fine sands with climbing ripple cross lamination (unit 1), middle varved silts (unit 2) and upper fine sands with climbing ripple cross lamination (unit 3).

Description

(a) Unit 1

The lower unit consists of 8 facies, which alternate one with another within the sedimentary column (Fig. 3). The thickness of the unit varies from 2 to 10 metres.

The fine sands and/or sandy silts with climbing ripples of type B (Jopling & Walker, 1968) predominate within the unit. They form beds from 0.5 up to 3.0 m thick. This facies shows fining-upward sequences from fine-sand to silty sand (graded beds). Within the thick sequences rippled fine sand (2.3-3.3phi) alternate frequently with

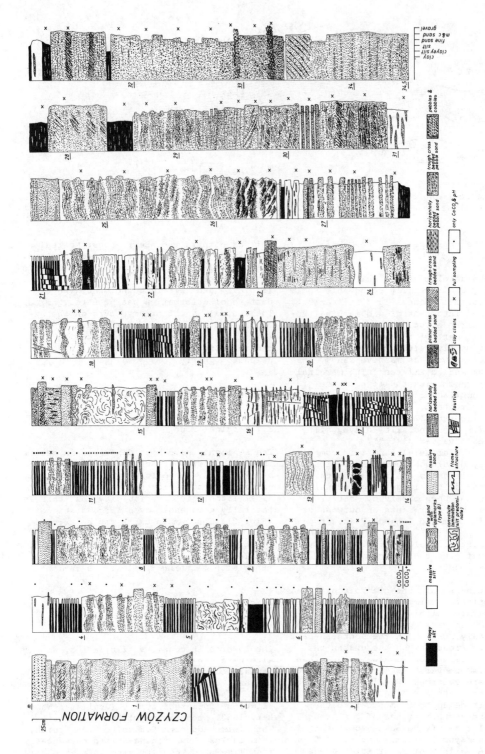

Fig. 3. Sedimentological log of the marginal glaciolacustrine sediments.
Location of the log is in Fig. 2.

the rippled silty sands (3.3-4.3phi). The latter may be massive, or indicate weakly rippled bedding (Fig. 3). The thickness of the individual graded beds is 10-20 cm.

Another rippled sands are represented by the small scale cross-bedded sets. The size of individual structures is usually 3-5 cm and the thickness of sets reaches up to 0.5 m.

The horizontally bedded fine and medium sands occur in the thin sets, of about 20-50 cm. They often alternate with the planar cross bedded medium sands. The latter appear in 10-30 cm thick sets. The horizontally and cross bedded sands form together individual beds up to 1.5 m thick. The trough cross-bedding occurs there only occasionally.

The massive sands are 10-30 cm thick structureless bodies of fine and medium sands. Their occurrence is usually laterally limited, as they appear as lenses within laminated silts and massive silts.

The laminated silts and sands occur in beds 0.5-1.0 m thick. These are formed of alternating, 1-3 cm thick laminae of massive or finely laminated silt and structureless sand. The silt units are usually thicker than the sands. The sands are fine to medium and are relatively well sorted. In turn, silty units are poorly sorted. The contact between the silts and sands is not sharp.

The massive silts are poorly sorted, brownish-grey silts (3.8-4.0phi) which occur in 10-40 cm thick beds, usually connected with the laminated silts and/or massive sands.

The massive clayey silts are poorly sorted, black clayey silts (4.0-4.5phi) which occur in 5-25 cm thick beds. Very often, there are observed thin sandy lenticles within the clayey silts. The facies is associated with the rippled or horizontally and cross bedded sand rarely with the laminated silts. The most important facies within unit 1 are rippled fine sands, horizontally and cross bedded sands and laminated silts. Other facies are rare and occur as thin beds.

(b) Unit 2

The middle unit consists 4 facies, which also alternate one with another (Fig. 3). The thickness of the unit varies from 20 m up to 30 m.

The most characteristic and most frequent facies within the unit are varved silts i.e. rhythmically laminated brownish-grey massive silts (3.8-4.8phi) and black clayey silts (4.5-5.5phi) (Fig. 4). The thickness of the individual laminae is usually 1-3 cm, although thicker laminae are common. The silt units are generally thicker than the black clayey silts, although occasionally clayey silts of greater thickness occur (Fig. 3 & 4). The silts usually exhibit gradual contacts with the black, clayey laminae, and the thicker silty units are finely laminated (Fig. 4). The thickness of the rhythmite beds varies from 0.3 m up to 2.0 m.

The other facies of unit 2 have the same characteristics as within unit 1. They are: laminated silts, massive sands and fine sands with climbing ripples of type B. The laminated silts are infrequent and form 20-50 cm thick beds within the rhythmites. Also, massive sand is infrequent and thin, occurring both within the rippled sands and rhythmites (Fig. 3). Two facies: silty rhythmites (varved silts) and rippled sands are most frequent and form 0.3-2.0 m thick beds which alternate one with another (Fig. 3 & 4). In addition, the sediments of unit 2 contain numerous flame structures, convolute beds and faults, that not occur within units 1 and 3.

(c) Unit 3

This unit is comprised of the same facies and facies associations as unit 1, although they are only partially exposed within sections. Unit 3 represents the uppermost glaciolacustrine sequence of the marginal zone, and this is often truncated by erosion.

Interpretation

The glaciolacustrine sediments of the marginal zone represent a continuous depositional sequence, as in all units climbing ripple drift is a dominant facies. Climbing ripples are associated with cross- and horizontally bedded sands and/or laminated silts in the lower and upper units, whereas in the middle unit they are associated with rhythmites.

The sediment was transported as bedload by density underflow currents on gently sloping delta foreset surfaces. The cross- and horizontally bedded sands may represent the distributary channels of the upper part of delta. In turn, the laminated silt was deposited between channels in quiet water from suspension. The massive silt and massive clayey silt were likely deposited in shallow water (Sturm, 1979; Sturm & Matter, 1978). The massive sands may represent grain flow deposits (Shaw, 1977) formed on a gently sloping delta front. Thus, unit 1 & 3 most probably represent the upper delta depositional environment. The occurrence of the rhythmites in unit 2 suggests a deeper basin. The alternation of sand and rhythmites is typical, however, of deposition in proximal, glaciolacustrine environments (Gustavson et al.,1975; Gustavson, 1975a). The silty-clayey silty rhythmites were probably deposited in a lake dominated by continuous underflows due to separation of sediments on the deltaic slope. These

Fig. 4. Deltaic rhythmites (varved silts) of the marginal zone
of the Kleszczow Graben.

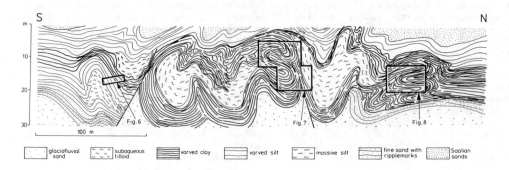

Fig. 5. The sediment succession in the central zone of the Elsterian
proglacial lake of the Kleszczow Graben. Location of the section is
in Fig. 1.

rhythmites have mostly thin, fine-grained couplets with some current structures. On the other hand, the multiple laminations may reflect rather short term fluctuations in sediment influx and disperal (Sturm & Matter, 1978; Ashley et al., 1985). The clay-poor lithology of rhythmites suggests, also that they were deposited in the proximal lower delta environment. Massive sands and convolute silty beds, which appear within the rhythmites represent, most pro-

bably, the slump induced deposits formed on the dalta slope (Shaw, 1977; Shaw & Archer, 1978; Leckie & McCann, 1982; Kelly & Martini 1986; Eyles et al., 1987). Moreover, numerous other gravity induced penecontemporaneous deformation horizons such as load casts and ball and pillow structures have been described from the Elsterian glaciolacustrine sediments in different parts of the Kleszczow Graben (Brodzikowski et al., 1987). Earthquakes resulting from graben tectonic activity have been suggested as a main trigger mechanism for their development.

3.2 Central zone

The sediments of the central zone are strongly deformed (Fig. 5). This deformation creates difficulty in recognizing complete sedimentary sequences. Hence, data have been collected in several short profiles. The glaciolacustrine suite described is about 30-40 m thick, and it indicates strong vertical and lateral lithological variability (Fig. 5). Five main units can be recognized, namely the lower fine sands with climbing ripples (unit 1), varved silts with massive sands and diamictons (unit 2), varved clays (unit 3), varved clays with diamictons (unit 4) and the uppermost massive and varved silts (unit 5).

Description

(a) Unit 1

The lower unit consists of the same facies as the unit 1 of the marginal zone, although fine sands with climbing ripples (type B) and massive silts predominate. The thickness of the unit varies from 2 to 6 metres.

(b) Unit 2

This unit consists of 3 facies. These are rhythmites which are comprised of 1-10 cm thick, alternating laminae of ripple drift fine sand (3.3-4.3phi), white, finely laminated clayey silt (4.5-5.5phi) and black, massive, clayey silt (4.5-6.3phi). Moreover, the layers of the massive sand (3.5-4.5phi) and diamictons occur within the rhythmites. The rhythmite beds are 30-50 cm thick and are separated one from the another by the 20-50 cm thick beds of the massive sands. The diamictons are massive or crudely stratified, matrix supported, muddy deposits which consists of small clasts up to 10 cm diameter as well as occasional large erratics up to 1.0 m diameter. The diamictons occur within the sedimentary sequences as 5-10 m thick bodies of limited lateral extent and they are strongly deformed. The other diamictons form thin layers within the rhythmite couplets (Fig. 6). The thickness of unit 2 varies from 6 up to 15 metres.

(c) Unit 3

The unit consists of two types of varved clays. The first, classical varves comprise the lower, white-grey silt bed (5.0-6.5phi) and the upper, dark grey to black clay bed (6.0-7.0phi). The thickness of a single varve varies from 5 to 40 cm and the light bed is usually two or more times thicker than the dark bed (Figs. 6, 7 & 8). There is no cor-

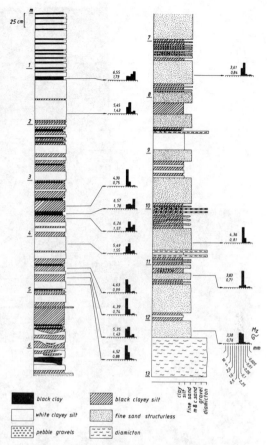

Fig. 6. The sedimentological log of the lake deposits from the central zone of the Kleszczow Graben. Mean size and sorting is in *phi* scale. Location of the log is in Fig. 5.

relation between clay layer thickness and silt layer thickness. The contact between the light and dark beds is usually sharp.

The next type of varved clays is comprised of a more complex lithological succession. One varve contains a 1-3 cm thick black clay (6.0-7.0phi) at the top, and alternating layers of the white clayey silt (4.5-5.5phi) and black clayey silt (4.5-6.5phi) at the bottom. The thickness of the bottom layer varies from 10 cm up to 1 m (Fig. 6). The contacts between clayey silts are gradual, whereas the boundary with the topmost clay is always sharp. The main difference between the two black layers within the single varve is that clay is poorly sorted (1.7-1.8phi), but the clayey silt is much better sorted (0.75-1.50phi). The thickness of unit 3 varies from 6 up to 15 metres.

Fig. 7. Varved clays interstratified with diamictons; central zone of the Kleszczow Graben. Location of the section is in Fig. 5.

Fig. 8. Typical varved clays from the central zone of the Kleszczow Graben. Location of the section is in the Fig. 5.

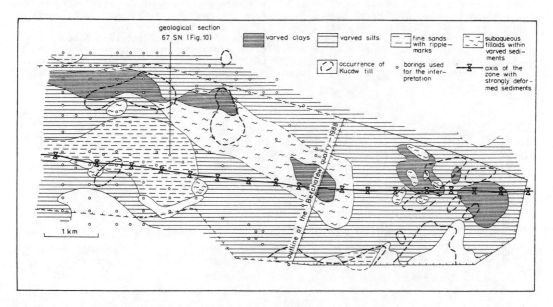

Fig. 9. Distribution of the Elsterian glaciolacustrine sediments in the Kleszczow Graben.

(d) Unit 4

This unit is comprised of the same facies as unit 3, and in addition contains large, iregular bodies of massive or crudely laminated diamictons (Figs. 5 & 7). Moreover, varved clays contain single dropstones up to 10 cm diameter. The thickness of the unit is about 20-25 m, including 10-15 m thick diamicton bodies.

(e) Unit 5

The uppermost unit consists of the same facies as the unit 2 of the marginal zone, although the massive silts may here reach up to 5 m.

Interpretation

The complete sequence of the central zone can be deduced from several sites. Unit 1 occurs always at the bottom part and unit 5 at the topmost part of the sequence. Units 2, 3 and 4 occur in the middle and they are laterally juxtaposed. Unit 2 occurs near the sediments of the marginal zone and unit 3 forms the most distal part of the basin. In places, varved clays of unit 3 overlie units 2 and 4, lying directly below the uppermost varved silts.

The lowermost and uppermost units represent the first and final, shallow basin stages of lake development. They are interpreted to be similar to the sedimentary units in the marginal zone. The varved clays of units 3 and 4 represent typical bottom facies of a deep, thermally stratified basin (Ashley, 1975; Smith et al., 1982). The rhythmites of unit 2 may represent the bottom sediments as they clearly show density underflow sedimentation (thin sandy layers within couplets). This and the fact that unit 2 occurs on the boundary between the central and marginal zones suggest that the unit is a lower delta slope deposit. The clay layers are, however, much finer than in the varved silts of the marginal zone. Hence, unit 2 was likely deposited on the lake bottom rather on the lowermost part of the delta.

Diamictons and massive sands are usually associated with gravity flow deposition in highly proximal environments of ice-contact lakes (Gustavson, 1975a, 1975b; Evenson et al., 1977; Kelly & Martini, 1986; Eyles, 1987; Eyles et al., 1987, 1988). Their occurrence within the distal varved clays is rather unusual, though Shaw & Archer (1978, 1979) described some massive sands within the winter layers. These were interpreted to be a result of winter turbidity current under stagnant-ice conditions, during deglaciation. Such interpretation is not appriopriate for sediments within the Kleszczow Graben because they were deposited during the ice advance.

4 DISTRIBUTION OF THE SEDIMENTS IN THE LAKE BASIN

The sections with the Early Pleistocene (Upper Elsterian) glaciolacustrine sediments were systematically investigated as they were exposed during excavation of Belchatow quarry. Also, boring records were used for the interpretation of the area lying west from the quarry (Krzyszkowski & Czerwonka, 1991). Fig. 9 presents a schematic distribution of the Upper Elsterian glaciolacustrine sediments in the Kleszczow Graben. It is obvious from the map that the bottom facies, namely varved clays occur sporadically

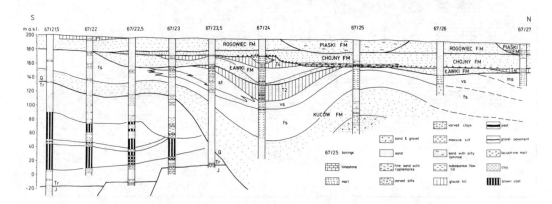

Fig. 10. Geological cross-section through trough sediments in the northern part of the Kleszczow Graben beyond the Belchatow quarry. Location of the section is in Fig. 9. fs - fine sands with climbing ripples, vs - varved silts, ms - massive silts, st - subaqueous tilloids interbedded with varved clays.

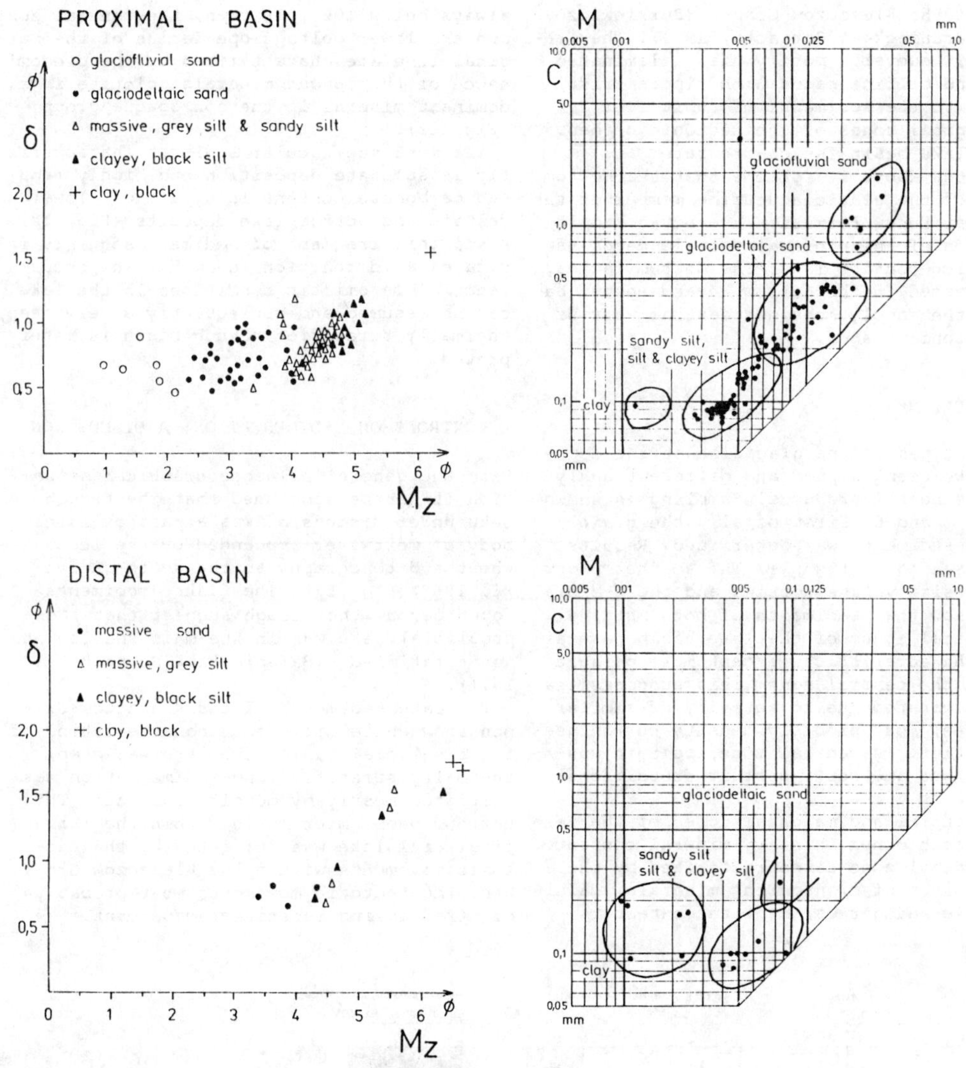
PROXIMAL BASIN
o glaciofluvial sand
• glaciodeltaic sand
Δ massive, grey silt & sandy silt
▲ clayey, black silt
+ clay, black
δ
ϕ
Mz
M
C
mm
glaciofluvial sand
glaciodeltaic sand
sandy silt,
silt & clayey silt
clay
DISTAL BASIN
• massive sand
Δ massive, grey silt
▲ clayey, black silt
+ clay, black
δ
ϕ
Mz
M
C
mm
glaciodeltaic sand
sandy silt
silt & clayey silt
clay

Fig. 11. Grain size characteristics of the glaciolacustrine sediments. Scatter plot of mean size versus sorting (left, grain size in phi) and C-M diagram (right, grain size in mm).

and only in the narrow, up to 1 km wide, zone in the axial part of the trough. The rhythmites with diamictons are more common and occur usually in the same zone, but generally on its margins. The marginal zones of the Kleszczow Graben are comprised of only varved silts and/or fine sands with ripple drift.

The axis of the deepest portion of the Elsterian glacial lake basin within the Kleszczow Graben (marked by the bottom facies) is coincident with the axis of the zone with strongest deformation of sediments. Only a part of Elsterian bottom facies is strongly deformed. The deformation occurs in the eastern part of the trough. These sediments in the western part of the trough are only gently deformed (Fig. 10). The deformation of the Elsterian sediments is assumed to be mostly post-depositional, from the Saalian times (Krzyszkowski, 1989a; Krzyszkowski & Czerwonka, 1991).

The area of the Kleszczow Graben which was examined is only 9 km long and 3-4 km wide (Fig. 9). The deposits described in the trough represent only a part of larger lake basin The Elsterian glaciolacustrine deposits were commonly described in areas north and south of the Kleszczow Graben (Jurkiewiczowa, 1961; Baraniecka & Sarnacka, 1971). They represent, however, mostly fine laminated silts and/or fine sands with ripple drift, similar to the sediments of unit 1 of the marginal zones of the tectonic graben. Bottom lake basin facies are rare.

From the above it follows that the bottom facies of the glaciolacustrine member of the Elsterian Kucow Formation are not a simple extention of the proximal deltaic deposits of the ice-contact lake. The bottom facies were created in the narrow, overdeepened basin of the deeply subsided central part of the tectonic graben.

5. MISCELLANEA

Several types of the glaciolacustrine deposits have been sampled and different analyses have been introduced. Sampling is shown in fig. 3 and 6. First of all, the grain size of sediments was determined. Results are presented in fig. 11. The scatter plots of mean size versus sorting and the C-M diagrams show that sediments of both marginal and central zones of the lake basin have similar characteristics, except more pelagic clay in the central zone. This suggests that they originated due to activity of similar processes, most probably density current sedimentation. In central zone, pelagic suspension was another important depositional factor.

The central and marginal zones of the lake differ each other in heavy mineral content. The marginal zone is characterized by the predominance of transparent minerals, mainly amphibole and garnet; biotite content is always below 10% . In turn, the central zone and the lower delta slope facies of the marginal zone are characterized by the predominance of the opaque minerals; biotite is a dominant mineral in the non-opaque group (Fig. 12).

The most spectacular sediment characteristic is sulphate deposition and highly reduced carbonate content in a part of lower deltaic and bottom lake deposits (Fig. 12). Also, this fragment of sediment sequence indicates acid reaction (pH < 7). The occurrence of meromictic conditions in the lake can be assumed and consequently a very deep, thermally stratified trough basin is interpreted.

6 CONTROLS ON SEDIMENTATION: A DISCUSSION

From the general palaeogeographical situation it can be concluded that the trough lake under discussion was a part of large body of meltwater impounded by the ice sheet and topography to the south (Rozycki, 1967)(Fig. 13). The lake sediments found beyond the trough suggest that the proglacial lake was in the main shallow and non-stratified (Baraniecka & Sarnacka, 1971).

The lake sediments of the Kleszczow Graben show quite different characteristics of the proglacial basin: the lake was deep, thermally stratified and sedimentation was initiated mostly by density currents. This narrow, deep microbasin within the large proglacial lake was initiated by the tectonic movements within the Kleszczow Graben. The tectonic movements most probably occurred during isostatic adjustment. The

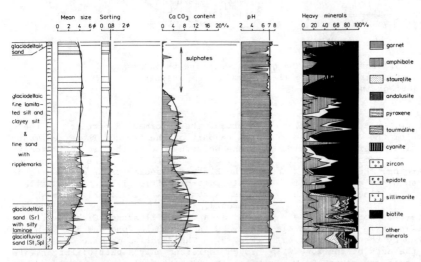

Fig. 12. The chemical and mineral characteristics of the Elsterian glaciolacustrine sediments in the Kleszczow Graben (Grain size in phi).

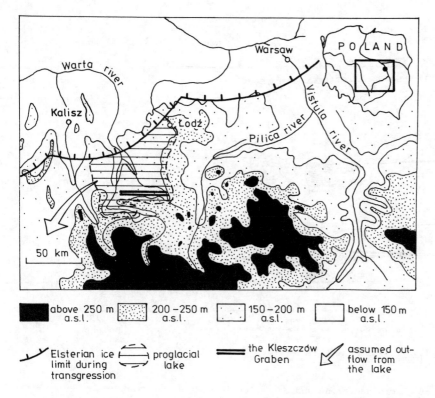

| above 250 m a.s.l. | 200 – 250 m a.s.l. | 150 – 200 m a.s.l. | below 150 m a.s.l. |

Elsterian ice limit during transgression | proglacial lake | the Kleszczów Graben | assumed out-flow from the lake

Fig. 13. A palaeogeographic map of the Upper Elsterian proglacial lake in central Poland.

increased subsidence of the crust was not compensated for by sediment influx and consequently a deep basin was created. The thermal stratification might have originated specifically in the overdeepened basin and/or occurred in the large part of the distal part of the proglacial lake (Fig. 13). The great depth of the basin is also well confirmed by the occurrence of meromictic conditions in the lake.

The sedimentary processes within the trough are not clearly associated with the processes in other parts of the proglacial lake, with the exception of the sediment influx from the ice-sheet and/or nonglacial rivers. The lake basin within the trough formed two zones: marginal and central. The margin was relatively shallow and gently inclined to the axis of the trough forming, most probably, gently dipping slopes. The continuous flow of the sediments near the bottom of the lake can be interpreted from the sedimentary structures. At the same time subaqueous slumps, which were likely generated by earthquakes, formed interflows in the thermally stratified water. The central zone of the lake basin was much deeper, because this was located in the second-order graben formed by faulting in the axial part of the tensional trough (Fig. 14). Thus steep slo-

pes (fault scarps) occurred within the basin and they have marked the boundary between the marginal and central zones. The grain and mass flows were generated on these slopes, forming thick massive sand bodies (Lowe, 1976; Shaw, 1977; Kelly & Martini, 1986; Eyles et al., 1987) and/or diamictons within the bottom rhythmites. The diamictons were probably transported as rafts within mass flows derived from the up slope failure of the older, extensive till units, which were exposed on the margins of the central zone. A glacial origin, for the large diamicton bodies is proposed because of their association with large, well rounded erratics (Krzyszkowski, 1989b). Hence, they should be named as the subaqueous tilloids (re-deposited tills). The term subaqueous flow tills should be left for proximal sediments (Evenson et al., 1977). Eyles et al. (1988) described large diamicton bodies, on steep valley slopes, which were derived by the repeated downslope movement and mixing of heterogeneous sediments including outwash gravels and glaciolacustrine deposits. They assume, that such deposits can be analogues for ancient sediments deposited on the margins of areas of active rifting. This model is appropriate for the Kleszczow Graben as the large diamicton bodies were

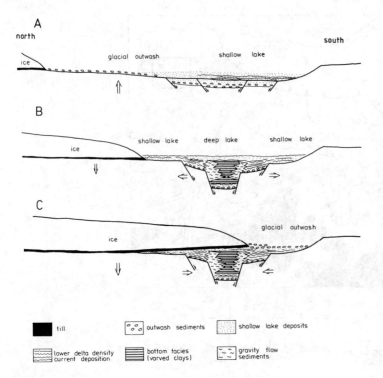

Fig. 14. A model of glaciolacustrine sedimentation in the tectonically active zone. A - initial stage, B - main stage, C - final stage. Detailed explanation in the text.

deposited on the margins of the tectonically active basin. The glacial origin for the diamictons in the Kleszczow Graben is, however, clear as the underlying glacial deposits are the only source of the large erratics (Scandinavian crystalline rocks). On the other hand, the thin layers of diamictons might have originated by the mixing of coarse and fine deposits during mass movement (Cohen, 1983; Eyles et al., 1987, 1988). The axis of the central, second-order graben is occupied only by bottom varved clays (Fig. 9 & 14). They were deposited in quiet areas reached only by pelagic suspension and an occasional density underflow as indicated from varve bedding and their grain size characteristics (Fig. 11). These are characterized by high clay content, both within the summer (up to 25%) and winter layers (up to 45%). Seasonal overturning as mechanism for bringing clay to the lake bottom (Sturm & Matter, 1978; Matthews, 1956) is not adequate in the meromictic lake. Probably, other explanations must be introduced (Smith et al. 1982).
A model of the proglacial lake sedimentation in the tectonically active basin is depicted in figure 14. When the ice-sheet was

far to the north proglacial outwash sediments were deposited proximally and more distally, shallow lake deposits (Fig. 14A). The main stage began, when rifting started in the tectonically active basin and sedimentation started in the small, deep lake (Fig. 14B). The final stage is characterized again by shallow lake and/or outwash deposition near the margin of the transgresing ice-sheet

7 CONCLUSIONS

It seems that besides the ice-contact and distal lakes the a type of the proglacial lake can be found - the tectonically influenced proglacial lake. This have many specific features. The tectonically influenced proglacial lake sediments are characterized by their abnormally great thickness (hundreds of metres) and their lateral association with proglacial deposits with quite opposite sedimentological characteristics. The tectonically influenced suites were deposited in very deep, in part meromictic basins. Hence, they are comprised of of similar sediments as the deep, thermally stratified, distal lakes. The specific characteristic here is that the bottom facies can be interstratified with the gravity flow sediments, mostly diamictons, which are usually connected with highly proximal environments. Within the tectonically influenced lakes such deposits are created on the subaqueous fault scarps during earthquakes.

It seems, that tectonically influenced proglacial lakes may be common in some glaciated areas, but they are not recognized satisfactorily, yet. On the other hand, recent and the Pleistocene tectonically influenced proglacial suites may be a good guide to recognize genesis of more ancient glacial suites.

ACKNOWLEDGEMENTS

Gail M. Ashley and William P. Warren have provided helpful discussion and improved the English. I am grateful also to J. A. Czerwonka for assistance in the field and laboratory.

REFERENCES

Ashley, G.M. 1975. Rhythmic sedimentation in glacial Lake Hitchcock, Massachusetts-Connecticut. In A.V. Jopling & B.C. McDonald (eds.) Glaciofluvial and Glaciolacustrine Sedimentation, SEPM Spec. Publ. 23, Tulsa, Oklahoma: 304-320.
Ashley, G.M. 1989. Classification of glaciolacustrine sediments. In R.P. Goldthwait & C.L. Matsch (eds.) Genetic Classification of Glacigenic Deposits, A.A. Balkema, Rotterdam: 243-260.
Ashley, G.M., Shaw, J. & Smith, N.D. 1985. Glacial Sedimentary Environments. SEPM Short Course 16, Tulsa, Oklahoma, 246 p.
Baraniecka, M.D. & Sarnacka, Z. 1971. Stratygrafia czwartorzedu i paleogeografia dorzecza Widawki. Biuletyn Instytutu Geologicznego 254: 157-270.
Brodzikowski, K., Haluszczak, A., Krzyszkowski, D. & Van Loon, A.J. 1987. Genesis and diagnostic value of large-scale gravity-induced penecontemporaneous deformation horizons in the Quaternary sediments of the Kleszczow Graben (central Poland). In M.E. Jones & R.M.F. Preston (eds.) Deformation of Sediments and Sedimentary Rocks, Geol. Soc. Spec. Publ. 29: 287-298.
Cohen, J.M. 1983. Subaquatic mass flows in a high energy ice marginal deltaic environment and problems with the identification of flow tills. In E.B. Evenson, Ch. Schlüchter & J. Rabassa (eds.) Tills and Related Deposits, A.A. Balkema, Rotterdam: 255-267.
Evenson, E.B., Dreimanis, A. & Newsome, J.W. 1977. Subaquatic flow tills: a new interpretation for the genesis of some lamina-ted till deposits. Boreas 6: 115-133.
Eyles, N. 1987. Late Pleistocene debris-flow deposits in large glacial lakes in British Columbia and Alaska. Sedimentary Geology 53: 33-71.
Eyles, N., Clark, B.M. & Clague, J.J. 1987. Coarse-grained sediment gravity flow facies in a large supraglacial lake. Sedimentology 34: 193-216.
Eyles, N., Eyles, C.H. & McCabe, A.M. 1988. Late Pleistocene Subaerial debris - flow facies of the Bow Valley, near Banff, Canadian Rocky Mountains. Sedimentology 35: 465-480.
Eyles, N. & Miall, A.D. 1984. Glacial Facies. In R.G. Walker (ed.) Facies models, 2nd ed., Geoscience Canada Reprint series 1: 15-38.
Gustavson, T.C. 1975a. Sedimentation and physical limnology in proglacial Malaspina Lake, southeastern Alaska. In A.V. Jopling & B.C. McDonald (eds.) Glaciofluvial and Glaciolacustrine Sedimentation, SEPM Spec. Publ. 23, Tulsa, Oklahoma: 249-263.
Gustavson, T.C. 1975b. Bathymetry and sediment distribution in proglacial Malaspina Lake, Alaska. Jour. Sed. Petrol. 45: 450-461.
Gustavson, T.C., Ashley, G.M. & Boothroyd, J.C. 1975. Depositional sequences in glaciolacustrine deltas. In A.V. Jopling & B.C. McDonald (eds.) Glaciofluvial and Glaciolacustrine Sedimentation, SEPM Spec. Publ. 23, Tulsa, Oklahoma: 264-280.
Jopling, A.V. & Walker, R.G. 1968. Morphology and origin of ripple-drift cross-laminations, with examples from the Pleistocene of Massachusetts. Jour. Sed. Petrol. 38: 971-984.
Jurkiewiczowa, I. 1961. Czwartorzed dorzecza Widawki. Biuletyn Instytutu Geologicznego 169: 175-206.
Kelly, R.I. & Martini, I.P. 1986. Pleistocene glacio-lacustrine deltaic deposits of the Scarborough Formation, Ontario, Canada. Sedimentary Geology 47: 27-52.
Krzyszkowski, D. 1989a. The tectonic deformation of Quaternary deposits within the Kleszczow Graben, central Poland. Tectonophysics 163: 285-287.
Krzyszkowski, D. 1989b. Quaternary Mixtites: Glacial and/or Nonglacial Deposition within the Kleszczow Graben, Central Poland. Abstacts, 28th Int. Geol. Congr., Washington, D.C., vol. 2: 233-234.
Krzyszkowski, D. 1991. Middle Pleistocene stratigraphy of Poland: a review. Proc. Geol. Ass. 102: 201-215.
Krzyszkowski, D. & Czerwonka, J.A. 1991. Quaternary Geology of the Kleszczow Graben (Central Poland): a study based on boreholes from the western forefield of the "Belchatow" outcrop. Quaternary Studies in Poland 11: in press.
Leckie, D.A. & McCann, S.B. 1982. Glaciolacustrine sedimentation on low slope

prograding delta. In R. Davison-Arnott, W. Nickling & B.D. Fahey (eds.) Proc. 6th Guelph Symp. on Geomorph., Res. in Glacial, Glaciofluvial and Glaciolacustrine Systems: 261-278.

Lowe, D.R. 1976. Grain flows and grain flow deposits. Jour. Sed. Petrol. 46: 188-199.

Mathews, W.H. 1956. Physical limnology and sedimentation in a glacial lake. Geol. Soc. Amer. Bull. 67: 537-552.

Miall, A.D. 1977. A review of the braided-river depositional environment. Earth Science Rev. 13: 1-62.

Miall, A.D. 1978. Lithofacies types and vertical profile models in braided river deposits: a summary. In A.D. Miall (ed.) Fluvial Sedimentology, Can. Soc. Petrol. Geol. Memoir 5, Calgary, Canada: 597-604.

Miall, A.D. 1984. Glaciofluvial transport and deposition. In N. Eyles (ed.) Glacial Geology, Pergamon Press, New York: 168-183.

Rozycki, S. Z. 1967. Plejstocen Polski Srodkowej. PWN Warszawa, p. 251.

Shaw, J. 1977. Sedimentation in an alpine lake during deglaciation, Okanagan Valley, British Columbia, Canada. Geografiska Annaler 59A: 221-240.

Shaw, J. & Archer, J. 1978. Winter turbidity deposits in Late Pleistocene glaciolacustrine varves, Okanagan Valley, British Columbia, Canada. Boreas 7: 123-130.

Shaw, J. & Archer, J. 1979. Deglaciation and glaciolacustrine conditions, Okanagan Valley, Brittish Columbia, Canada. In Ch. Schlüchter (ed.) Moraines and varves, A.A. Balkema, Rotterdam: 347-355.

Smith, N.D. 1971. Transverse bars and braiding in the Lower Platte River, Nebraska. Geol. Soc. Am. Bull. 82: 3407-3420.

Smith, N.D. 1974. Sedimentology and bar formation in the upper Kicking Horse River, a braided outwash stream. Jour. Geol. 81: 205-223.

Smith, N.D., Venol, M.A. & Kennedy, S.K. 1982. Comparison of sedimentation regimes in four glacier-fed lakes of Western Alberta. In R. Davison-Arnott, W. Nickling & B.D. Fahey (eds.) Proc. 6th Guelph Symp. on Geomorph., Res. in Glacial, Glaciofluvial and Glaciolacustrine systems: 203-238.

Sturm, M. 1979. Origin and composition of clastic varves. In Ch. Schlüchter (ed.) Moraines and Varves, A.A. Balkema, Rotterdam: 281-285.

Sturm, M. & Matter, A. 1978. Turbidities and varves in Lake Brienz (Switzerland): deposition of clastic detritus by density currents. In A. Matter & M.E. Tucker (eds.) Modern and Ancient Lake Sediments, Int. Ass. Sedim. Spec. Publ. 2: 147-168.

Formation and Deformation of Glacial Deposits, Warren & Croot (eds) © 1994 Balkema, Rotterdam, ISBN 90 5410 096 6

Tunnel-valley fans of the St. Croix moraine, east-central Minnesota, USA

Carrie J. Patterson
Minnesota Geological Survey, St. Paul, Minn., USA

ABSTRACT: Tunnel valleys are deeply incised, broad, subglacial drainageways. They were a significant component of the subglacial drainage system of the Late Wisconsinan Superior lobe in central Minnesota. The valleys form a radial pattern that parallels the inferred flow lines for the lobe. The surface trenches that represent the former tunnel-valley courses are 500–2300 m wide, 5–45 km long, and, although partially infilled, 10–25 m deep. Fans formed at the ice margin at the mouths of the tunnel valleys.

More than twenty-five tunnel-valley-fan systems of the Superior lobe have been mapped in Minnesota and western Wisconsin. Five are described in this paper. The sediment at the mouths of the tunnel valleys forms a roughly fan-shaped deposit that can be highly pitted. The fans have a complex ice-contact stratigraphy and consist mainly of sand and gravel. They are as much as 46 km^2 in area and can project 65 m above the surrounding landscape. In the Elk River fan complex, the gravel attributed to the fan is 50–80 m thick.

Tunnel-valley-fan systems developed at all major stillstands of the Superior lobe. The height of the fans suggests that the water that deposited the sand and gravel was flowing under a significant hydraulic head. The size and spacing of the tunnel valleys imply that the subglacial drainage system was adequately developed and prevented excessive pore water pressures and ice instability from developing.

1 INTRODUCTION

In 1956 Wright mapped "drainageways" in the area that was covered by the Superior lobe (Wright, 1956). Cushing (1963; Wright et al., 1964) later interpreted these features to be subglacial channels similar to those described for northern Europe (e.g., Grube, 1983; Woodland, 1970). Finally, Wright (1973) called the features tunnel valleys and described a radial network of channels that extended from the southern tip of Lake Superior to the Mississippi River in east-central Minnesota (Fig. 1). More recent mappers of Superior lobe deposits have further elaborated upon and extended Wright's initial mapping (Johnson, 1986, in prep.; Lehr, 1991; Mooers, 1989; Savina et al., 1979).

Wright (1973) hypothesized that the tunnel valleys of the Superior lobe were created during catastrophic discharges of stored, basal meltwater when the ice was at its maximum. He thought that a frozen margin would inhibit the evacuation of subglacial meltwater, but that the water would burst through the frozen zone when pressures became high enough. Mooers (1989) objected to this theory on the grounds that more water was needed to create the tunnel valleys than could be produced by basal melt. He hypothesized that the tunnel valleys formed not as an integrated system when the Superior lobe was at its maximum and the ice was presumably thick and cold-based, but in discrete segments at retreat positions when the ice was probably warm-based and thin enough to allow surface meltwater to reach the bed. He also thought that bank-full conditions may never have been reached in these valleys.

During recent mapping of the deposits of the Superior lobe of the Laurentide Ice Sheet in east-central Minnesota, I noted the

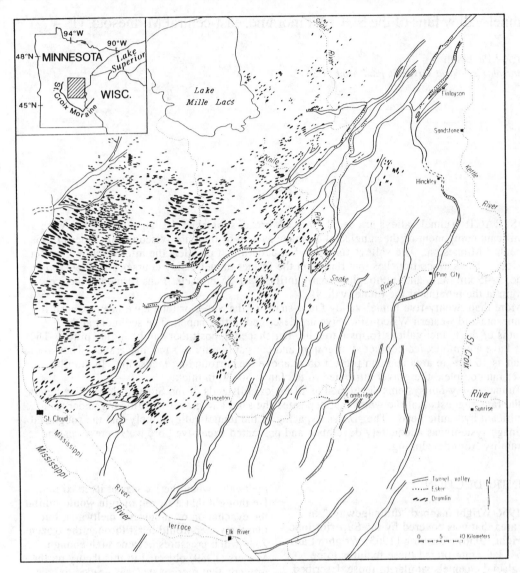

Figure 1. The distribution of drumlins, tunnel valleys, and eskers of the Superior lobe in east-central Minnesota. From Wright (1972) except for location map.

presence of large sand and gravel deposits at the mouths of tunnel valleys. A broader look at the region outside the area of my initial mapping confirmed that these deposits are found along the entire southern sector of the area once covered by the Superior lobe. In this paper I describe the three-dimensional geometry and lithostratigraphy of five tunnel-valley-fan systems and discuss their likely mode of origin. This proposed model reflects

the present understanding of the history of deglaciation in Minnesota and western Wisconsin, which is based on detailed mapping of surficial deposits.

2 STUDY AREA

The study area encompasses ten counties in east-central Minnesota and two counties in

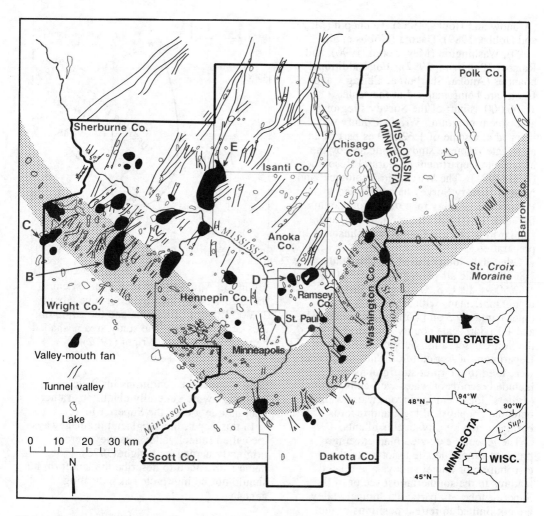

Figure 2. Tunnel valleys, valley-mouth fans, and the St. Croix moraine in the study area of east-central Minnesota and adjoining counties in Wisconsin. Lettered features are landforms discussed in the text: A, Comfort Lake tunnel valley and fan; B, Buffalo Lake tunnel valley and fan; C, Goose Lake tunnel valley and fan; D, Arsenal fan and Lino Lakes tunnel-valley system; and E, Elk River fan complex.

western Wisconsin that contain the ice-marginal deposits of the Late Wisconsinan Superior lobe. Three large rivers flow through the area; the Mississippi and Minnesota Rivers flow from the northwest and southwest, respectively, to their confluence in the Minneapolis–St. Paul (Twin Cities) Metropolitan Area (Fig. 2). The Twin Cities are situated in a topographic low. The St. Croix River in the study area forms the Wisconsin-Minnesota border and joins the

Mississippi River 50 km downstream from the confluence of the Minnesota and the Mississippi Rivers.

2.1 Sources of data

Surficial deposits of five counties in the study area (Fig. 2) were mapped at the scale 1:100,000 as part of the county atlas program of the Minnesota Geological Survey: Scott

71

(Aronow and Hobbs, 1982), Hennepin (Meyer and Hobbs, 1989), Dakota (Hobbs et al., 1990), Washington (Meyer et al, 1990), and Ramsey (Patterson, 1992). Four additional counties (Anoka, Sherburne, Chisago, and Isanti) are being mapped at the scale 1:200,000 as part of the Survey's regional assessment program. Wright County was mapped at a scale of 1:62,500 as part of an aggregate-resource study conducted by the Minnesota Department of Natural Resources (Lehr, 1991). The Wisconsin Geological and Natural History Survey has published a Pleistocene geology map of Barron County at the scale 1:100,000 (Johnson, 1986), and a similar map of Polk County by Johnson is in preparation.

Other data sources used for this study were 1:80,000-scale air photographs, 1:24,000- and 1:62,500-scale U.S. Geological Survey topographic maps with 10 or 20 ft (approximately 3 and 6 m) contour intervals, and 1:15,840-scale county soil surveys from the Soil Conservation Service, U.S. Department of Agriculture. Subsurface data were used to interpret stratigraphy and included records of water-well construction (drillers' logs), cuttings from water wells, borehole geophysical logs, and records of test borings made by private consultants.

The recently extended map coverage presents a more detailed picture of the distribution of tunnel valleys and related features in the southernmost sector of the Superior lobe. In particular, tunnel valleys are not limited to retreat positions of the Superior lobe. Linear collapsed areas, which represent buried tunnel valleys, extend to the Mississippi River, crossing it in places, and also breach the St. Croix moraine, as suggested by Wright (1973). Smaller, buried tunnel valleys are found beyond the St. Croix moraine as far as the terminal extra-morainic position (Fig. 2). In addition, large fans, which formed at the mouths of the tunnel valleys, are responsible for much of the debris that defines the St. Croix moraine and lesser moraines of the Superior lobe. Some of the highest parts of the St. Croix moraine are sand and gravel deposits of this origin.

Tunnel valleys and fans formed at the maximum position of the Superior lobe and continued to form along the retreating ice front. The persistence of these features

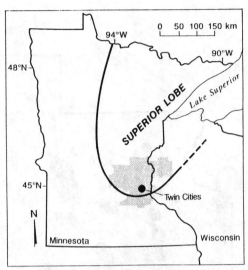

Figure 3. Approximate maximum extent of the Late Wisconsinan Superior lobe, about 20,000 B.P., in Minnesota and the adjoining area of Wisconsin. The study area is shaded gray. Adapted from Wright (1972).

suggests that the conditions for their formation were generally ubiquitous rather than unique within the Superior lobe.

In this paper, the subglacial drainageways are called tunnel valleys, as they were originally named by Wright (1973). The name is intended to describe the landform but should not be interpreted as indicating genesis.

2.2 St. Croix phase of the Superior lobe

The Superior lobe in its St. Croix phase reached its maximum in Minnesota (Fig. 3) about 20,000 years ago (Wright, 1972). This ice lobe entered Minnesota from the north through the Lake Superior lowland, flowed into a structural basin in east-central Minnesota called the Twin Cities lowland (Fig. 3), and built the prominent St. Croix moraine (often considered the terminal position) 220 km to the southwest of the southern tip of Lake Superior. There were probably advances as much as 15 km beyond the St. Croix moraine (Johnson and Savina, 1987), but the St. Croix moraine is the largest moraine complex formed by the Superior lobe

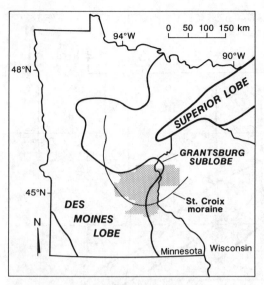

Figure 4. Positions of the advancing
northwestern-source Des Moines lobe and
Grantsburg sublobe, and the retreating
Superior lobe at about 14,000 B.P. The study
area is shaded gray. Adapted from Wright
(1972).

and represents a stable, near-maximum
position of the lobe.

The northeast-trending Twin Cities
lowland, which is composed of Paleozoic
sedimentary rocks, and especially Cambrian
sandstone, underlies most of the glacial
sediment in the study area. Northeast of the
study area, and extending into the Lake
Superior lowland, Middle Proterozoic
sandstone underlies Superior lobe deposits.

A gray till, which is more clayey than the
red, sandy, Superior lobe till, is sometimes
encountered in deep drill holes, indicating
that the Superior lobe in the study area
probably advanced over a patchy distribution
of pre–Late Wisconsinan glacial sediment.
Deep valleys, which have been interpreted as
interglacial stream courses, had incised this
older till before the advance of the Superior
lobe.

The St. Croix moraine in the Twin Cities
area of Minneapolis and St. Paul is a broad,
hummocky, lake-dotted zone, 10 to 30 km
wide. The Mississippi River generally
follows the inside arc of the moraine until it
flows through a gap in north Minneapolis

(Meyer et al., 1990). The moraine is "one of
the most sharply defined glacial features in
the Great Lakes region" (Wright, 1972, p.
527). However, in most places within the
study area it is buried by the glacial sediment
of a later advance of a northwestern-source
ice lobe, the Grantsburg sublobe of the Des
Moines lobe (Fig. 4). Because the deposits of
this later lobe are generally less than 15 m
thick, the character of the deposits of the St.
Croix phase can still be discerned.

2.3 Tunnel valleys

The tunnel valleys in the area of ice-lobe
overlap have been partially infilled with
glacial sediment of the Grantsburg sublobe.
The trend and approximate dimensions of
many segments of the tunnel valleys were
preserved, an indication that they were
apparently filled with ice at the time of the
later, Grantsburg advance. The linear
collapsed areas that are inferred to represent
the tunnel-valley segments are 500–2300 m
wide (averaging 1200 m), 10–25 m deep, and
5–45 km long. Wright (1973) gives the
dimensions of the Superior lobe tunnel
valleys as 180–1000 m wide (averaging 300
m) and as much as 30 m deep (averaging 10
m). The range of dimensions of the collapsed
areas is consistent with the tunnel valleys
described by Wright (1973).

The outermost tunnel-valley segments are
shorter, narrower, and more closely spaced
than those to the northeast. In addition, their
associated fans are typically smaller and of
lower relief. The orientations of the
outermost tier of tunnel valleys in the study
area define an arc of at least 100 degrees.
The arcs formed by the inner tiers of tunnel
valleys decrease progressively northeastward
to 80 and eventually 30 degrees. These inner-
tier valleys are wider, deeper, and have larger
fans. The narrowing of the arc of tunnel-
valley fanning represents a change in the
geometry of the active ice during retreat. The
lobate shape is replaced by a more fingerlike
projection of ice, similar in shape to the phase
of the Superior lobe at about 14,000 B.P. (Fig.
4).

Most of the tunnel valleys of the Superior
lobe are in areas of sedimentary bedrock. The
tunnels continue a short distance into the area

73

Figure 5A. Comfort Lake fan and associated tunnel valleys. See Figure 2 for location. From U.S. Geological Survey 1:62,500-scale Forest Lake quadrangle (contour interval 20 ft, approximately 6 m).

underlain by granite, which is in most places deeply weathered. In areas of Minnesota where crystalline rock is covered by sediment of the Superior lobe, drumlins rather than tunnel valleys are the dominant subglacial landform.

2.4 Valley-mouth fans

The valley-mouth deposits, located where the subglacial channels discharge at the ice front, are roughly fan-shaped and usually highly pitted. They project as much as 65 m above the surrounding landscape. The extent of individual fans, which is determined on the basis of surface expression, is as much as 46 sq km. Subsurface data show that the fan deposits are as much as 80 m thick. The sediment exposed at the surface is predominantly sand and gravel with cobbles and boulders. The fan deposits are an enthusiastically mined aggregate resource where they are not deeply buried by sediment of the Grantsburg sublobe. Some of the fans in the Twin Cities area of Minneapolis and St.

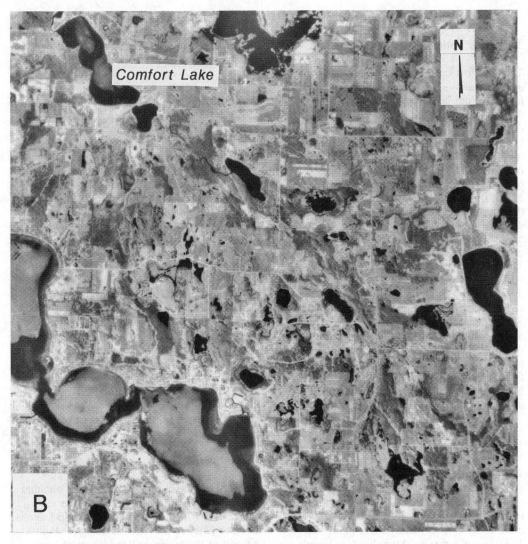

Figure 5B. Air photo of the Comfort Lake fan and associated tunnel valleys (Mark Hurd Aerial Surveys, EKA 938, 4-9-77).

Paul no longer have coherent surface expression owing to extensive mining in the past.

3 DESCRIPTION AND INTERPRETATION

This section contains descriptions of five tunnel and fan systems that are characteristic of the landforms found throughout the study area. The locations of specific fans are shown on Figure 2. The tunnels valleys were mapped by delineating the elongate depressions that remain at the surface. The fans were recognized and mapped by their geomorphology and surficial geology. Because all of the fans and tunnels in this area have been partially buried by later glacial deposits, these mapping methods probably underestimate the lateral extent of the fan deposits and the depth of the tunnels.

3.1 Comfort Lake tunnel valley and fan

3.1.1 Description

The Comfort Lake tunnel valley (Fig. 5) is approximately one kilometer wide at the fan mouth. Comfort Lake, a water-filled depression within the tunnel valley, is 12 m deep. Although buried by deposits of the Grantsburg sublobe, the fan retains its topographic expression, owing to its position near the Grantsburg margin where the overlapping glacial sediment is thin. An extension of the tunnel valley is discernible through the fan as a series of topographic depressions and lakes that probably formed by collapse from ice-block melt-out. The surface of the fan is dotted with numerous small pits. There also is a faint linear pattern of low-relief ridges and trenches (9–15 m high and 30–120 m wide) that cover the fan and are aligned with the direction of meltwater flow. Subsurface information is abundant. There are more than 80 drillers' logs for the area of the fan. Cuttings, descriptive logs, and a gamma log, all from a well southeast of Cranberry Lake, show that deposits of Superior lobe sand and gravel are approximately 25 m thick at this locality.

3.1.2 Interpretation

The Comfort Lake tunnel valley is a segment of a larger tunnel-valley system that extends through much of the St. Croix moraine and is inferred to cross the St. Croix River, a distance of approximately 30 km (Fig. 2). The channel contains throughout its length many eskers which postdate the initial tunnel-valley downcutting. The tunnel valley system is broken up into segments 5 km long with fans at the mouths of at least three segments. The Comfort Lake fan is the northernmost fan and the last to form during ice retreat.

The episodic nature of ice retreat and fan formation is inferred from the 5-km spacing of the fans. The tunnel valleys feeding the fans may have been only a few kilometers long but eroded up-ice as the ice front retreated.

The pits that dot the surface of the Comfort Lake fan probably formed from the melting of either blocks of ice that were discharged with

the water and sediment, or from stagnant ice beneath the fan. The linear pattern of ridges and trenches on the fan resembles the pattern created by channels on modern alluvial fans.

3.2 Buffalo Lake tunnel valley and fan

3.2.1 Description

The fan west of Buffalo Lake (Fig. 6) has an area of approximately 25 km^2, with a maximum relief of 30 m. However, the land surface west of the fan is as high as, and in places higher than, the fan itself. The fan lies within the area traditionally mapped as the St. Croix moraine (Fig. 2). Subsurface information is contained in 20 drillers' logs available for the area. There are small tunnel segments beyond the moraine position (Fig. 2). Three tunnel valleys trend toward the Buffalo Lake fan. The eastern, Lake Pulaski branch is 27 m deep and 1300 m wide. The two western channels, the Sullivan Lake and Light Foot Lake branches, are much narrower and shallower. Well logs for the area record gravelly units 2–45 m thick, with an average thickness of 15 m. This fan also displays the linear pattern of low-relief trenches and ridges that are present on the Comfort Lake fan.

3.2.2 Interpretation

The Lake Pulaski branch is an older channel that contributed little sediment to the main part of the Buffalo Lake fan. Buffalo Lake is a water-filled depression in the Lake Pulaski branch of the tunnel valley; it was at least partially ice-filled at the time of deposition of the fan. The geomorphology of the fan indicates that most of the sediment was received from the two smaller tunnel valleys.

3.3 Goose Lake tunnel valley and fan

3.3.1 Description

The Goose Lake tunnel valley (Fig. 7) is 3 to 25 m deep and 1 to 1.5 km wide near the apex of the fan. The fan covers 30 km^2 and rises 60 m above the surrounding landscape and 85 m above the bottom of the tunnel

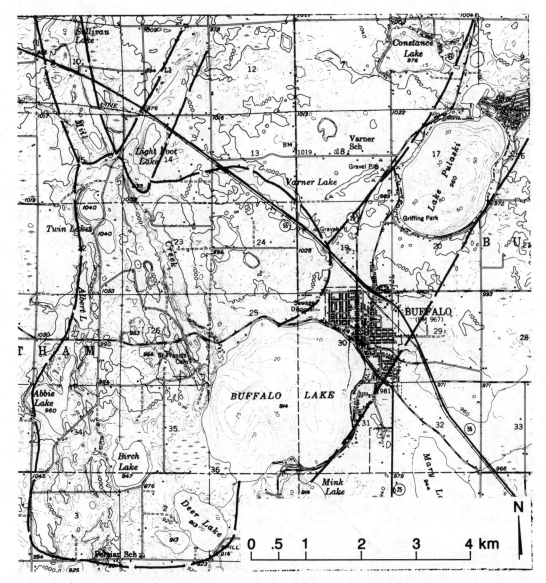

Figure 6. Buffalo Lake fan and associated tunnel valleys. See Figure 2 for location. From U.S. Geological Survey 1:62,500-scale Buffalo and Monticello quadrangles (contour interval 20 ft, approximately 6 m).

valley. The fan is within the area that has been called the St. Croix moraine (Fig. 2) and is the highest part of the moraine in Wright County (410 m or 1234 ft above sea level). Subsurface information for this fan is sparse. Information from eight well logs indicate a sand-and-gravel unit beneath approximately 20 m of diamicton of the Grantsburg sublobe.

Because no wells continue through this unit, the maximum thickness is unknown but is at least 14 m.

3.3.2 Interpretation

The Goose Lake fan is the outermost fan of

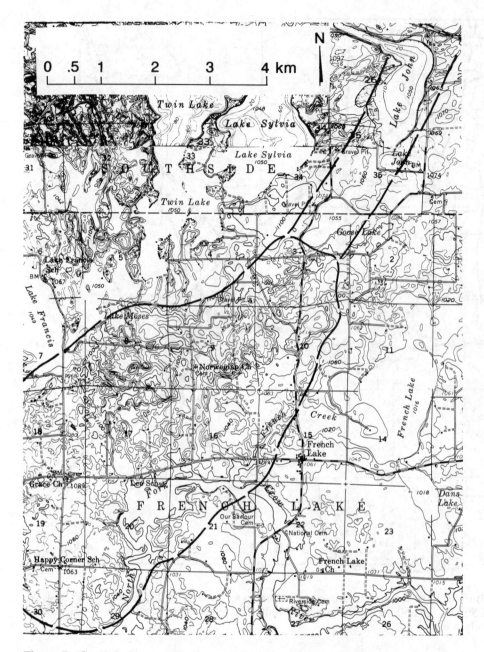

Figure 7. Goose Lake tunnel valley and fan. See Figure 2 for location. From U.S. Geological Survey 1:62,500-scale Annandale (contour interval 10 ft, approximately 3 m) and Cokato and Dassel (contour interval 20 ft, approximately 6 m) quadrangles.

this tunnel valley system (Fig. 2). The tunnel valley is traceable in the up-ice direction (northeast) for a distance of approximately 25 km. There are no other major fan deposits along this tunnel valley system; however, breaks in the trace of the tunnel valley may represent stable ice-front positions.

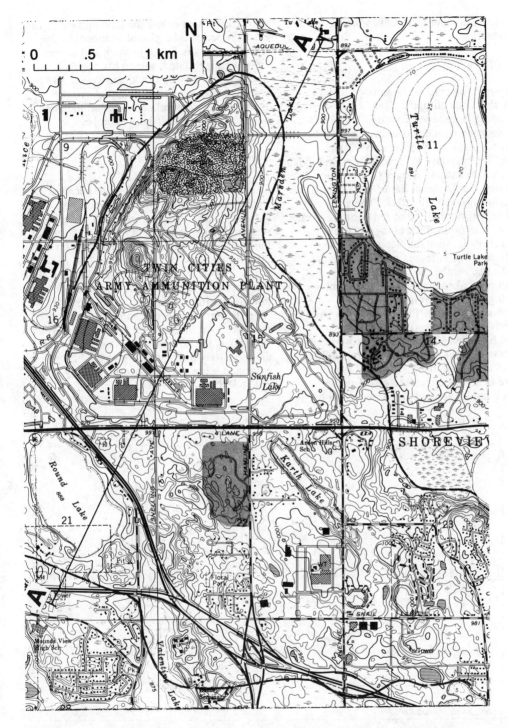

Figure 8. Arsenal fan. See Figure 2 for location. From U.S. Geological Survey 7.5-minute New Brighton quadrangle (contour interval 10 ft, approximately 3 m).

3.4 Arsenal fan and Lino Lakes tunnel-valley system

3.4.1 Description

The Arsenal fan (Fig. 8) is an impressive feature that rises abruptly 60 m above the surrounding landscape. Extensive subsurface study has been conducted in this area (G.N. Meyer, Minn. Geol. Survey, unpub. data, 1986–87) in response to problems of groundwater contamination. Meyer (1987) recognized the "Arsenal kame" as a deposit of a Superior lobe tunnel valley. The sand-and-gravel unit attributed to the fan-forming event is as much as 60 m thick along the cross section (Fig. 9). The fan deposit contains thin, discontinuous lenses of till, which are not indicated on the cross section.

The Lino Lakes tunnel-valley system, extending northeast from this fan for approximately 30 km (Fig. 2), is 4 km wide near the Arsenal fan and consists of two parallel branches. The linear collapsed areas that demarcate the tunnel valley are shallow, no more than 10 m deep. Bedrock lies 30 m or more beneath the surface of the tunnel valleys near the apex of the Arsenal fan but is only 10 to 25 m below the land surface to the east. Superior lobe glacial sediment in the tunnel valleys is 5 to 8 m thick and ranges from a consolidated, somewhat clayey diamicton to a sorted sand and gravel. The rest of the material within and above the tunnel valley is unrelated till from the later advance of the Grantsburg sublobe.

3.4.2 Interpretation

This fan is small in area (9 km^2) because the flanks have been removed by meltwater (Meyer, 1987). The gravelly unit in the fan may be thicker than the 60 m that the logs of borings indicate; it may attain a thickness of 90 m to the west, where the fan is 30 m higher. Now-separated gravel bodies that lie to the southwest are probably remnants of the fan. In addition, a fan deposit to the east (Fig. 2) may have been fed by the same drainage system (the Lino Lakes tunnel-valley system) as the Arsenal fan. The subglacial drainage may have eroded into the bedrock or exploited a pre-existing bedrock valley (Jirsa et al., 1986).

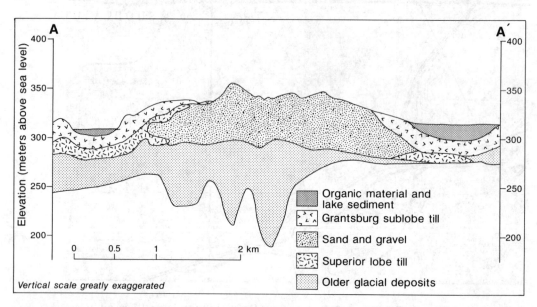

Figure 9. Cross section of the Arsenal fan. See Figure 8 for location of section line. Adapted from an unpublished interpretation by G.N. Meyer, Minn. Geol. Survey, 1986–87.

3.5 Elk River fan complex

3.5.1 Description

This fan complex is the largest in the study area (Fig. 2); sand-and-gravel deposits extend over 20 km and cover approximately 270 km^2. The most recently formed fan of this sequence, the Blue Lake fan, at the north end of the complex, remains distinct (Fig. 10). The fan is roughly 46 km^2 in area, has a maximum relief of 50 m, and extends 60 m above the floor of the Blue Lake tunnel valley. More than 50 drillers' logs are available for this area. The cross section runs the length of the fan (Fig. 11). Thick gravel-and-sand units are interrupted in places by till lenses, which are too small to be shown in the cross sections.

The Blue Lake tunnel valley is approximately one kilometer wide where it joins the northern part of the fan. It extends approximately 45 km to the northeast. Blue Lake, a water-filled depression within the tunnel valley, is as much as 10 m deep.

3.5.2 Interpretation

Many separate drainage events are represented in the Elk River fan complex. At least three main subglacial drainage systems fed the area, including the Blue Lake tunnel valley (Fig. 10) and two earlier valleys that enter the fan farther south and trend more to the west (these earlier valleys are indicated on Figure 2). The fan complex may even extend across the Mississippi River; a sand and gravel body immediately south of the uppermost terrace of the Mississippi River lines up with the Elk River deposit. The Elk River fan complex, the longest one in the study area, represents a system of tunnel valleys that discharged sediment over a longer period of time than any other in the study area.

The Blue Lake fan at the northern end of the Elk River complex was deposited by the outflow of one main tunnel valley, the Blue Lake tunnel valley, which had two distributary channels (Fig. 10). The longer, westernmost channel was occupied first, as indicated by the superposition of fans. Thick gravel deposits in the western channel imply that it became clogged with debris and the flow diverted to the shorter, eastern channel.

Much of the western portion of the Elk River fan complex has been reworked by the Grantsburg sublobe. In fact, the indicator bedrock erratic for the Grantsburg sublobe is the soft Pierre Shale, which crops out in western Minnesota and eastern North Dakota over 200 km west and northwest of the study area. It is found to a depth of at least 10 m in the fan. The bulk of the sand and gravel of the fan, however, was deposited by the Superior lobe. The Grantsburg sublobe contributed little material, but its meltwater reworked the deposit enough to contaminate it with shale and leave an altered surface expression.

4 DISCUSSION

The subglacial drainage of the Superior lobe, which persisted throughout the retreat of the lobe from the terminal position, was channeled through a system of tunnel valleys. Tunnel valleys and associated fans are far more numerous near the margin of the Superior lobe than previously recognized. The outermost (extra-morainic) tunnels and fans were the earliest to form. They are smaller than those that developed in and behind the St. Croix moraine, because the extra-morainic ice advances were short-lived and the meltwater volumes were smaller. These extra-morainic advances did not create recognizable moraines.

Eskers commonly occupy the tunnel valleys and are probably the product of diminished water fluxes and resulting ice closure (Boulton and Hindmarsh, 1987; Mooers, 1989). The eskers within the tunnel valleys of the Superior lobe are much smaller (30–150 m wide and generally less than 10 m high) than the tunnel valleys and much

Figure 10 (the following two pages). Elk River fan complex. See Figure 2 for location. A. From U.S. Geological Survey 1:62,500-scale Elk River (contour interval 20 ft, approximately 6 m) and St. Francis (contour interval 10 ft, approximately 3 m) quadrangles. B. Air photo of the same area (Mark Hurd Aerial Surveys, EKA-745, 4-9-77).

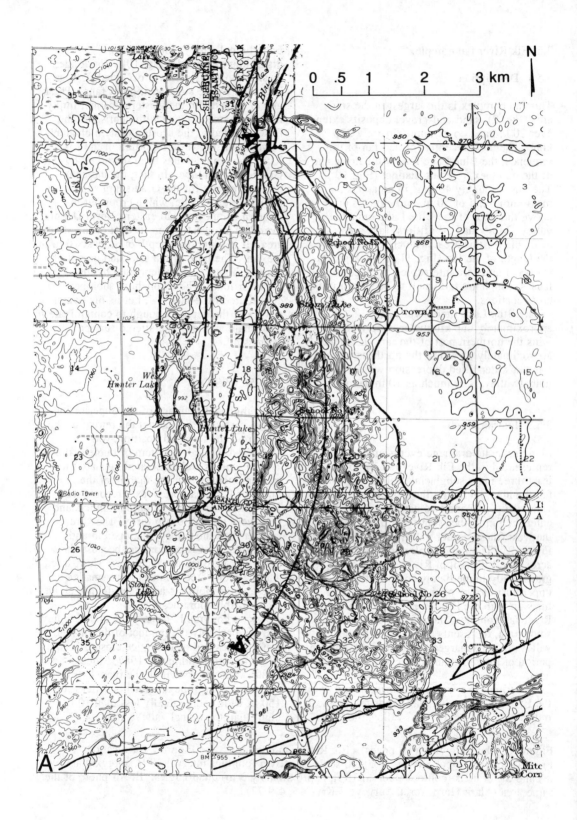

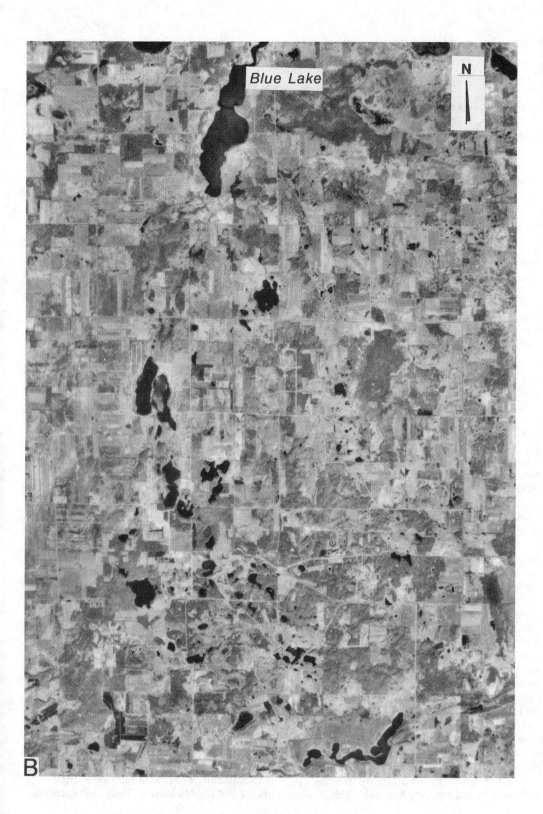

Blue Lake

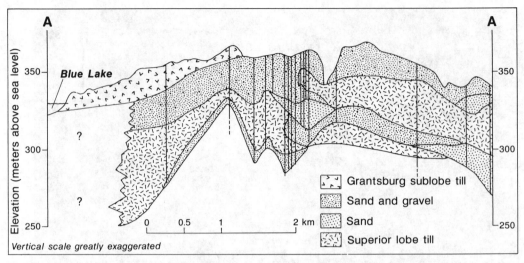

Figure 11. Cross section of the Elk River fan. See Figure 10A for the location of section line. The section line is curved and has not been projected onto a straight line. Therefore, distances between data points are preserved. Vertical lines on cross section indicate locations of wells from which subsurface information was obtained.

narrower than the apexes of the tunnel-valley fans.

Although there is evidence of a frozen margin in Minnesota (Mooers, 1989) and Wisconsin (Attig et al., 1989) when the Superior lobe was at its maximum, it does not necessarily follow that the frozen margin persisted throughout retreat, even if conditions remained cold enough for permafrost to exist. Some theories of tunnel-valley formation require a frozen margin to store meltwater (Wright, 1973; Attig et al., 1989). A frozen margin would create the necessary water pressures and hydraulic head needed to form tunnel valleys, but the probability of one existing whenever tunnel valleys formed is small for two reasons: (1) it takes time to redevelop a frozen margin at a retreat position, because the ice is no longer resting on a frozen substrate (a thawed zone develops up-ice because of pressure melting, and during retreat it becomes the substrate beneath the margin), and (2) tunnel valleys formed during the Pleistocene in areas where the ice margin was fronted by a large body of water (Gorrell and Shaw, 1991); they also exist in modern maritime glacial systems where a frozen margin is impossible (Gustavson and Boothroyd, 1982, 1987).

The tunnel valleys are primarily in areas underlain by sedimentary bedrock. The correspondence of tunnel valleys to permeable bedrock is contrary to the predictions of Boulton and Hindmarsh (1987), who suggested that tunnel-valley development and spacing is determined by the hydraulic transmissivity of the bed and that tunnel valleys develop over poor subglacial aquifers. Erodibility of sedimentary rock, rather than its hydraulic transmissivity, may be more important to the development and location of tunnel valleys.

A few tunnel valleys in the study area display first-order tributaries, but the majority of them are short, single channels. Most commonly seen are first-order distributaries. Such distributaries probably formed as one channel was choked by the buildup of gravel and ice closure as the flow of water diminished. The distributary channels were not occupied simultaneously. The pattern of widespread distributary channels and rare tributaries indicates that the drainageways were short-lived. The straight, nondendritic nature of the channels also suggests that the flux of water involved was large.

The sizes of the drainage events that deposited the fans can presumably be estimated from the sedimentology of the fans. The surface exposures that were mapped are predominantly sand, with relatively small gravel and cobble lenses. They are typical of subaerial alluvial fan deposits. The thick gravel sequences in the subsurface were once present at the surface but are now mined out. The boulders left behind in gravel pits are impressive in size and number but unfortunately cannot be used to reconstruct the depositional history of the fans because they are not in place. Detailed work on the sedimentology of the fans remains to be done.

The approximate dimensions of the tunnel valleys in the study area were preserved in the substrate where the valleys were filled with ice. The overall dimensions of the channels must be determined in order to model the formation of the tunnel valleys. Does the height of the fans give an indication of the extent to which the channel eroded into the

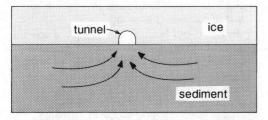

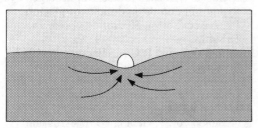

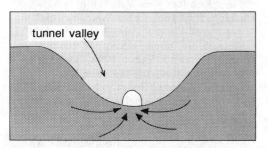

Figure 12. The origin of sediment-floored subglacial tunnels and tunnel valleys, as presented by Boulton and Hindmarsh (1987). Sediment creeps into the tunnel, an area of low pressure, and is carried away by meltwater. Modified from Boulton and Hindmarsh (1987)

ice? In the model of tunnel-valley formation offered by Boulton and Hindmarsh (1987; Fig. 12), the tunnel is very small compared to the size of the valley in the substrate. Such a small ice tunnel seems incapable of depositing the fans of the Superior lobe, which in places rise 85 m above the tunnel-valley floors. In addition, these fans may have been even higher when initially deposited if they were deposited on stagnant ice or contained large blocks of ice. The dimensions of the fans lead to the following alternative hypotheses: (1) the channels that deposited the fan sand and gravel were approximately 85 m high; or (2) the hydraulic head in the ice was at least 85 m high, and the velocity of the discharging fluid had a vertical component; or (3) the gradient of the subglacial valley rose rapidly as it approached the terminus and the discharge was englacial or supraglacial when it deposited the fans; or (4) the fans were deposited gradually as the tunnels melted upward through the ice.

The first hypothesis is the least likely: the flux of water required to fill a channel that large would be difficult to maintain. The modes of formation suggested in the other three hypotheses have been observed in modern glacial systems. There are ice-marginal water spouts or fountains in Greenland and Spitsbergen, Norway (Chamberlin and Salisbury, 1909, p. 281; Baranowski, 1973), and marginal fountains and sub- and englacial meltwater tunnels are the predominant means of meltwater discharge from the Malaspina Glacier in Alaska (Gustavson and Boothroyd, 1982, 1987). Although the Malaspina Glacier has been used as an analogue to the southeastern margin of the Laurentide Ice Sheet (Gustavson and Boothroyd, 1987), the tunnels and fans of the Malaspina system are as much as an order of magnitude smaller than those of the Late Wisconsinan Superior lobe. The flux of both water and sediment must have been much larger to form the features of the Superior lobe.

The apparent persistence of the Superior lobe at the St. Croix moraine, the lack of landforms associated with surging and an ensuing stagnation, and the steady retreat of the ice from the St. Croix moraine, as indicated by recessional ice positions, suggest that whatever the nature of the subglacial drainage, it supported stable ice movement. For the most part, the subglacial water pressures of the Superior lobe did not exceed ice-overburden pressures. Rather, the meltwater either drained away to a subglacial aquifer or was stored within or beneath the ice. On a regular basis, short-lived subglacial drainage channels evacuated excessive meltwater to maintain the stability of the ice mass. These drainageways deposited discrete fans at the ice margin. As the ice retreated, drainages were reopened along the same or similar trends because of perturbation in the bed surface or basal ice.

5 CONCLUSIONS

Tunnel valleys dominated the subglacial drainage system of the Superior lobe at its maximum and throughout steady retreat. The location and approximate trend of a tunnel valley, once established, persisted throughout retreat. The remains of the tunnel valleys are short, straight channels with few tributaries. They are 500–2300 m wide, 5–45 km long (although only a few kilometers may have been active at any one time) and at least 10–25 m deep.

The water that flowed through these tunnels deposited fans at the ice margin that are up to 85 m higher than tunnel floors. Individual fans are as much as 46 km^2 in area. The fans are commonly discrete features at ice-front positions but can be fan complexes that received more continuous deposition throughout retreat (i.e., the Elk River fan complex). The heights of the fans above the tunnel-valley floor indicate that the discharging subglacial water flowed under a high hydraulic head. A modern analogue for the drainage system of the Superior lobe may be the marginal fountains or water spouts that issue from sub- and englacial tunnels in glaciers in Alaska, Spitsbergen, and Greenland.

6 ACKNOWLEDGEMENTS

I thank Lee Clayton and John Attig of the Wisconsin Geological and Natural History Survey for helpful discussions, and Mark Johnson of Gustavus Adolphus College, St. Peter, Minnesota, and J.D. Lehr of the Minnesota Department of Natural Resources for making their unpublished maps available to me. I would also like to thank the reviewer and editor for helpful suggestions.

Partial funding for the geologic mapping in Anoka and Ramsey Counties was supported by Ramsey County and by an appropriation to the Minnesota Department of Natural Resources, Division of Waters, pursuant to the Minnesota Groundwater Protection Act of 1989.

REFERENCES

Attig, J.W., D.M. Mickelson, and L. Clayton. 1989. Late Wisconsin landform distribution and glacier-bed conditions in Wisconsin. Sed. Geol. 62: 399–405.

Aronow, S. and H.C. Hobbs. 1982. Surficial geologic map. In N.H. Balaban & P.L. McSwiggen (eds), Geologic atlas of Scott County, Minnesota. Minn. Geol. Survey County Atlas Ser. No. C-1, 1:100,000.

Baranowski, S. 1973. Geyser-like water spouts at Werenskioldbreen, Spitsbergen. In Symposium on the hydrology of glaciers: Int. Assoc. Scient. Hydrol. Pub. No. 95: 131–133.

Boulton, G.S. and R.C.A. Hindmarsh. 1987. Sediment deformation beneath glaciers: rheology and geological consequences. J. Geophy. Res. 92, No. B9: 9059–9082.

Chamberlin, T.C., and R.P. Salisbury, 1909. Geology, vol. 1, Geologic Processes and Their Results. New York, Holt.

Cushing, E.J. 1963. Late-Wisconsin pollen stratigraphy in east-central Minnesota. PhD thesis, Univ. Minn., USA. 165 p.

Gorrell, G., and J. Shaw. 1991. Deposition in an esker, bead and fan complex, Lanark, Ontario, Canada. Sed. Geol. 72: 285–314.

Grube, F. 1983. Tunnel valleys. In Ehlers, J. (ed.), Glacial deposits in north-west Europe. Rotterdam, Balkema: 257–258.

Gustavson, T.C. and J.C. Boothroyd. 1982. Subglacial fluvial erosion: a major source of stratified drift, Malaspina Glacier, Alaska. In Davidson-Arnott, R., Nickling, W., and Fahey, B.D. (eds), Research in glacial, glacial-fluvial and glacio-lacustrine systems. Guelph Sym. Geomorph., 6th,

1980, Proc.: Norwich, Conn., USA, Geo Books, p. 93–116.

Gustavson, T.C. and J.C. Boothroyd. 1987. A depositional model for outwash, sediment sources and hydrologic characteristics, Malaspina Glacier, Alaska: a modern analog of the southeastern margin of the Laurentide Ice Sheet. Geol. Soc. Am. Bull. 99: 187–200.

Hobbs, H.C., S. Aronow and C.J. Patterson. 1990. Surficial geology. In N.H. Balaban & H.C. Hobbs (eds) Geologic atlas of Dakota County, Minnesota. Minn. Geol. Surv. County Atlas Ser. No. C-6, 1:100,000.

Jirsa, M.A., B.B. Olsen and B.B. Bloomgren. 1986. Bedrock topographic map of the seven-county Twin Cities Metropolitan Area, Minnesota. Minn. Geol. Sur. Misc. Map Ser. No. M-55, 1:125,000.

Johnson, M.D. 1986. Pleistocene geology of Barron County, Wisconsin. Wisc. Geol. & Nat. Hist. Sur. Info. Circ. No. 55.

Johnson, M.D. In prep. Pleistocene geology of Polk County, Wisconsin. Wisc. Geol. & Nat. Hist. Sur. Info. Circ.

Johnson, M.D. and M. Savina. 1987. The late Wisconsin southern margin of the Superior lobe was 10 to 15 kilometers south of the St. Croix moraine. Geol. Soc. Am. Abs. with Prog. 19: 206.

Lehr, J.D. 1991. Aggregate study of Wright County. Min. Dept. Nat. Res., Minerals Div. Rept. No. 294, 18 p., map.

Meyer, G.N.. 1987. St. Croix moraine kames: debouchment deposits of subglacial tunnel valleys. Geol. Soc. Am. Abs. with Prog. 19: 233–234

Meyer, G.N. and H.C. Hobbs. 1989, Surficial geology. In N.H. Balaban (ed.), Geologic atlas of Hennepin County, Minnesota. Minn. Geol. Sur. County Atlas Ser. No. C-4, 1:100,000.

Meyer, G.N., R.W. Baker and C.J. Patterson. 1990, Surficial geology. In L. Swanson & G.N. Meyer (eds), Geologic atlas of Washington County, Minnesota. Minn. Geol. Surv. County Atlas Ser. No. C-5, 1:100,00.

Mooers, H.D. 1989. On the formation of the tunnel valleys of the Superior lobe, central Minnesota. Quat. Res. 32: 24–35.

Patterson, C.J. 1992. Surficial geology. In Swanson, L., & Meyer, G.N. (eds), Geologic atlas of Ramsey County, Minnesota. Minn. Geol. Sur. County Atlas Ser. No. C-7, 1:48,000.

Savina, M., R. Jacobson and D. Rodgers. 1979. Outwash deposits of central Dakota County, Minnesota. Unpub. rept., Dept. of Geology, Carleton College, Northfield, Minn.

Woodland, A.W. 1970. The buried tunnel valleys of East Anglia. Yorkshire Geol. Soc. Proc. 37: 521–578.

Wright, H.E., Jr. 1956. Glacial geology of eastern Minnesota. In G.M. Schwartz (ed.), Glacial geology, eastern Minnesota. Geol. Soc. Am., Ann. Meeting, Guidebook for field trip No. 3: 97–107.

Wright, H.E., Jr. 1972. Quaternary history of Minnesota. In P.K. Sims & G.B. Morey (eds), Geology of Minnesota: a centennial volume. Minn. Geol. Sur.: 515–547.

Wright, H.E., Jr. 1973. Tunnel valleys, glacial surges, and subglacial hydrology of the Superior Lobe, Minnesota. Geol. Soc. Am. Mem. No. 136: 251–276.

Wright, H.E., Jr., E.J. Cushing and R.G. Baker. 1964. Eastern Minnesota. Midwest Friends of the Pleistocene, 15th Ann. Field Conf.: 32 p.

Formation and Deformation of Glacial Deposits, Warren & Croot (eds) © 1994 Balkema, Rotterdam, ISBN 90 5410 096 6

The subglacially engorged eskers in the Lutto river basin, northeastern Finnish Lapland

Peter Johansson
Geological Survey of Finland

ABSTRACT: The subglacially engorged eskers formed in the Lutto river basin during the deglaciation phase, approximately 9400 years BP. Running down the slopes as low ridges they reflect the route of subglacial streams which found their way through fractures to the base of the ice sheet and carved tunnels leading downslope. The engorged eskers are sedimentologically complex. They are often build up of sand and gravel at the core and covered by a crust of till. In many cases grain size and degree of sorting are extremely variable even in different parts of the same ridge.

INTRODUCTION

Twenty-eight low ridge-like landforms were found in connection with the regional mapping of Quaternary deposits in 1988 and 1989 in the Lutto river basin in Finnish Lapland (Fig. 1). More detailed investigations showed these ridges to be subglacially engorged eskers which were first described in detail by Mannerfelt in Sweden (slukås in Swedish).

The purpose of the study was to make an inventory of the engorged eskers in the area and to study their stratigraphy and grain size distribution as well as their morphology and orientation in relation to the glacier flow. In the location and inventory of the engorged eskers aerial photographs, both black-and-white and infrared colour ones, were used. The new network of provisional roads required by the timber felling made it possible to dig cuts and test pits into the eskers by tractor excavator for the purpose of sampling and stratigraphical investigations.

The Lutto river basin is the largest known area of occurence of engorged eskers in Finland. Ridges resembling these have formerly been described by Tanner (1915, p. 319) and by Penttilä (1963, p. 54) from the Saariselkä region and by Porola (1981, p. 22-23) from the western side of Lake Inari. The greatest number of engorged eskers have been observed in the fell region of Sweden. Jämtland in Central Sweden is considered the type area, where Mannerfelt (1945) and later Lundqvist (1969)

have made detailed studies of the engorged eskers of Oviksfjällen and Sylälvsdalen. They have also been found south of Jämtland, in Värmland (Lundqvist 1958) and in northern Sweden (Frödin 1914 and Lundqvist 1987, p. 34). In central Norway engorged eskers exist, e.g. in Tröndelag (Mannerfelt 1945, p. 215 and Embleton and King 1975, p. 371).

Outside of Fennoscandia observations of engorged eskers are scarce. Stone (1959) and Sissons (1961) have studied similar esker-like landforms in the Southern Uplands of Scotland. In the arctic parts of Canada, ridges similar to engorged eskers have been observed. Bird (1967) refers to them as "valley eskers".

GLACIAL HISTORY

The Lutto river basin consists of low, forested hills, where the highest tops, rising over 400 m above sea level, are unforested or clad by low fell birches. The relative differences in altitude are 150-300 metres. The Lutto river flows eastwards in the middle of the region. It joins the Tuloma river and empties into the Arctic Ocean at the city of Murmansk.

The Lutto river basin was freed from the cover of the continental ice sheet approximately 9500 - 9400 years ago (Johansson 1988). The margin of the ice sheet receded in the study area to the southwest, towards the ice divide zone, which was located about 100 - 150 km from the study

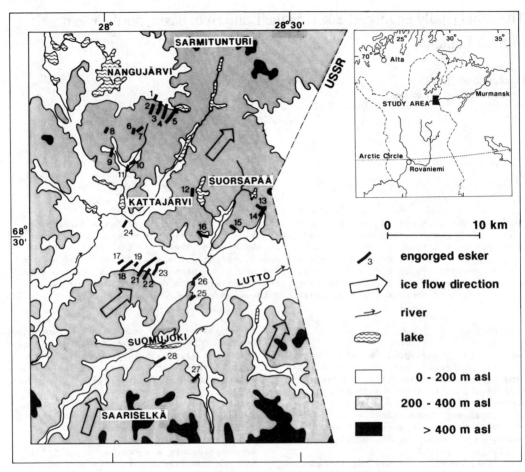

Fig. 1. Map of the location of the engorged eskers in the study area

Fig. 2. The engorged esker on the northern slope of Mademorosto (no. 5 in Fig 1) consists of a sequence of discontinous ridges and hummocks.

area. In the beginning of the deglaciation
when the continuous ice margin was still
tens of kilometres away to the northeast
of the study area, the highest tops were
starting to emerge as nunataks in the
middle of the glacier. Later, when the
glacier grew thinner and the margin receded to the study area, glacier lobes
were formed on the valley floors, e.g. in
the Lutto river basin. During the last
phase the lobes melted or turned into
fields of stagnant ice, finally melting in
the valleys (cf. Penttilä 1963, Saarnisto
1973).

GLACIOFLUVIAL LANDFORMS IN THE STUDY AREA

The deglaciation was accompanied by strong
meltwater flow. In the study area there
are several different types of glaciofluvial erosional and depositional landforms.
The depositional landforms may be classified according to their place of deposition
as inframarginal, marginal and extramar-
ginal (cf. Lundqvist 1979). Among the inframarginal formations especially the subglacial and englacial eskers are common in
the area. They are tens of kilometres long,
5 - 20 metres high ridges that consist of
well sorted glaciofluvial material. The
esker sequences lie in a southwest-north-
east direction, reflecting the network of
long glacial rivers that once existed in
the area. The topography of the area is
varied and the esker sequences tend to
follow the fracture zones in the bedrock
that lie in the direction of the latest
glacier flow. In the Saariselkä water-
divide zone there are also eskers that
cross fell ranges, showing that the meltwater was flowing in the ice tunnel under
strong hydrostatic pressure.

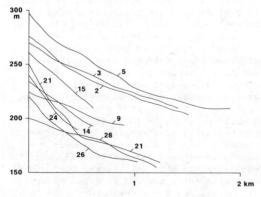

Fig. 3. Longitudinal sections of engorged
eskers. Numbers refer to the map in Fig.1.

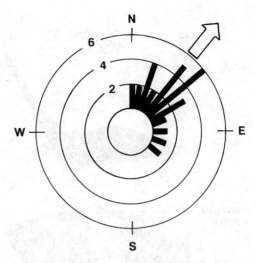

Fig. 4. Diagram showing the orientation of
the engorged eskers in relation to their
number. The white arrow shows the direction of glacier flow in the study area.

ENGORGRD ESKERS AND THEIR FORMATION

The engorged eskers are also inframarginal
formations. They differ from the eskers
proper described before as to their mode
of formation, size and material. As to
their morphology, the engorged eskers of
the Lutto river basin are steep-sided and
narrow ridges (Fig. 2). Their length is
between 100 metres and nearly three kilometres. The average height is 3 - 5 metres
and the highest are nearly 10 metres. The
engorged eskers wind gently down the con-
cave part of the slope (Fig. 3). The gradient of their base is between 1:7 and 1:40.
The ridges can be found either isolated
or in groups of several parallel ones. The
orientation of the ridges seems to be
mainly determined by the direction of the
slope. Most of them, however, have been
formed on a base that slopes toward the
northeast, the direction of the latest
glacier flow (Fig. 4).

The engorged eskers are supra-aquatic
landforms deposited by meltwaters flowing
from the glacier during the deglaciation.
Since some of the engorged eskers are cut
by lateral meltwater channels, they must,
however, have been deposited before these
channels were formed, i.e., before the
slope emerged from under the ice. Thus the
engorged eskers were formed during the initial stage of the deglaciation, when the
first hilltops emerged as nunataks from
under the thinning ice sheet (Fig. 5A).
The slopes and valleys around the nunataks

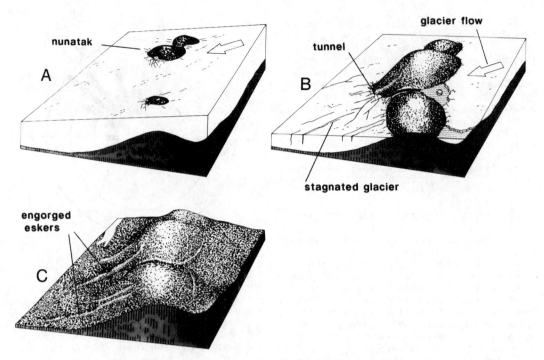

Fig. 5. Graphic presentation of the genesis of engorged eskers.

were then still covered by a hundreds of metres thick glacier. Meltwater flowing from the glacier collected around the nunatak, forming a marginal ice-lake. The dark rock absorbed the rays of the sun to a higher degree than the ice sheet, increasing the melting of the ice in its close vicinity. The meltwater made its way under the glacier along the fractures in the ice, opening meltwater tunnels down the slope. Fractures were formed especially in places where the glacier had already stagnated. Stagnation occured mainly on the northeastern slopes, since they were located on the distal side relative to the direction of glacier flow (Fig. 5B).

The meltwater transported debris from the ice margin and from the slope into the tunnels. The water flow washed and sorted the debris into different fractions. Current-bedded sand and gravel was deposited in the tunnel with the beds sloping either to the sides of the esker, or in the direction of the ridge downslope. Deposition and erosion alternated in the tunnel owing to changes in the water flow. The depositional process seems to have been interrupted occassionally, and the material already deposited on the floor of the tunnel was partly eroded, apparently to be redeposited downstream. Intercalations of fine sand and silt can also be

seen in the cuts. They were deposited when the flow of water became slower or stopped altogether as a consequence of seasonal changes or clogging of the tunnel. The clogging was caused by debris earlier deposited on the floor of the tunnel or by ice blocks that had fallen down, forming a dam. As the hydraulic head increased the pressure against the walls of the tunnel and against the dam increased. This led to the collapse of the dam, and so the meltwater broke through to a lower level, widening the tunnel and transporting with it material already sedimented. The formation of an engorged esker was a complex process with various changes and disturbances during the sedimentation.

The orientation and gentle winding of an engorged esker down the slope reflects the subglacial tunnel in the base of the glacier. The engorged eskers resemble subglacial chutes, erosional features corresponding to the engorged eskers and frequently existing in their vicinity. When the overlying glacier grew thinner, the tunnel turned into an open channel. The supraglacial material was lowered onto the sediments. In this way the engorged eskers were frequently covered by a bed of till, also including big stones and blocks (cf. Mannerfelt 1945, p. 124).

At Kattajärvi (no. 11 in Fig. 1) an

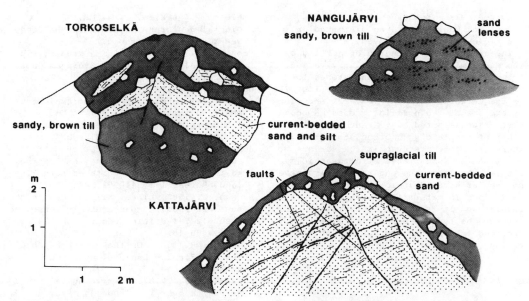

Fig. 6. The stratigraphy of engorged eskers in three different cuts.

engorged esker consisting of current-bedded sand is covered by a 20 - 80 cm thick till bed (Fig. 6). The structures of the sand show distinct faults, formed under load. They were formed as the till was lowered onto the frozen sand layers. Till was mixed with sorted material also by being loosened from the tunnel floors (cf. Kleman 1988, p. 43). In cuts dug into the engorged eskers till has been encountered also inside the ridges, at their sides and base. In the cut, east of Nangujärvi (no. 3), the till was loose and stony, containing several horizontal sandy lenses. At Torkoselkä (no. 15) the core of the esker was unstructured basal till. Overlying it there were sandy and silty layers, sloping towards the sides of the esker. Uppermost there was loose, stony till (Fig. 6).

Some of the small, less than two metres high engorged eskers consist of only till, as the small ridges east of Nangujärvi (no. 4) and near the river Lutto (no. 26). Their till contains elongated stones with a clear orientation, the same as that of the ridge. Some of the ridges consisting of only till may have formed as water-logged debris was squeezed into a subglacial hollow or fracture, as described by Kujansuu (1967) and Haselton (1979). The reason for the squeezing would have been the differences in pressure between the fracture and the surrounding glacier. Consequently the orientation of the stones would have formed as a result of squeezing or during the flow of the debris along the subglacial fracture downhill, and not as a result of the active movement of the glacier.

No faults or disturbances formed after deposition and caused by the movements of the glacier have been encountered in any of the cuts made into the engorged eskers. The glacier would have had to turn passive already before the deposition of the

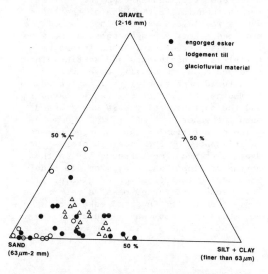

Fig. 7. Ternary diagram showing grain size distribution of engorged eskers compared with that of lodgement till and glacio-fluvial material in surrounding areas.

93

engorged esker, because otherwise the structures would have been disturbed or the ridges totally destroyed.

The grain-size and degree of sorting of the esker material is irregularly varied, from sandy till with abundant stones to well-sorted gravel, sand and silt (Fig. 7). In the networks formed by several engorged eskers one of two parallel ridges may consist of well-sorted gravel and sand, the other of completely unsorted stony till. There are contrasts even in the various parts of the same ridge. Stratigraphic studies show that the engorged eskers have no regular stratigraphic sequence. The variations in material are completely irregular and it is almost impossible to determine the type of deposit forming an engorged esker on the mere basis of morphology.

SUMMARY AND CONCLUSIONS

The largest area of occurence of subglacially engorged eskers in Finland is situated in the Lutto river basin. Twenty-eight ridges have been found in this area. Their morphology and stratigraphy shows a close resemblance to the engorged eskers observed in Sweden.

They formed during the initial stage of the deglaciation, when the first hilltops emerged as nunataks and the slopes around them were still covered by the glacier. At the margin of the stagnant glacier crevasses appeared, and in them the meltwater formed tunnels. At the base of the tunnel debris was laid down, forming a subglacial engorged esker. Rapid changes in the water flow caused great variations in the material as to grain-size and sorting. The engorged esker is usually covered by unstructured supraglacial till, which was formed as the roof of the tunnel caved in and the supraglacial debris was lowered onto the esker.

The orientation of the engorged eskers is influenced by the local landforms. However, they occur most frequently on the northeast slope, which was the lee side considering the direction of glacier flow. The glacier had turned passive before the deposition of the engorged esker, because no faults or disturbances caused by the movements of the glacier have been encountered.

REFERENCES

Bird, J. 1967. The Physiography of Arctic Canada. Baltimore: Johns Hopkins Press.

Embleton, C. & King, C. A. M..1975. Glacial geomorphology. London, Halsted, New York: Edward Arnold.

Frödin, J. 1914. Geografiska studier vid St. Luleälvs källområde. Sveriges Geol. Unders. Ser. C 257.

Haselton, G. M. 1979. Some glaciogenic landforms in Glacier Bay National Monument, Southeastern Alaska. In Ch. Schlüchter (ed.). Moraines and Varves, p. 197-205. Rotterdam: Balkema.

Johansson, P. 1988. Deglaciation pattern and ice-dammed lakes along the Saariselkä mountain range in north-eastern Finland. Boreas 17: 541-552.

Kleman, J. 1988. Linear till ridges in the southern Swedish mountains. Evidence for a subglacial origin. Geogr. Ann. 70 A (1-2): 35-45.

Kujansuu, R. 1967. On the deglaciation of western Finnish Lapland. Bull. Comm. géol. Finlande 232.

Lundqvist, J. 1958. Beskrivning till jordartskarta över Värmlandslän. Sveriges Geol. Unders. Ser. Ca 38.

Lundqvist, J. 1969. Beskrivning till jordartskarta över Jämtlandslän. (Summary: Description to the map of the Quaternary deposits of the County of Jämtland). Sveriges Geol. Unders. Ser. Ca 45.

Lundqvist, J. 1979. Morphogenetic classification of glaciofluvial deposits. Sveriges Geol. Unders. Ser. C 767.

Lundqvist, J. 1987. Beskrivning till jordartskarta över Västernorrlandslän och förutvarande Fällsjö K:n (Summary: Description of the map of the Quaternary deposits of the County of Västernorrland with northern Ångermanland). Sveriges Geol. Unders. Ser. Ca 55.

Mannerfelt, C. M:son 1945. Några glacialmorfologiska formelement. Geogr. Ann.27.

Penttilä, S. 1963. The deglaciation of the Laanila area, Finnish Lapland. Bull. Comm. géol. Finlande 203.

Porola, P. 1981. Deglasiaatio Koarvikoddsin alueella Inarin länsiosassa. Unpubblished M. Sc. thesis. Institute of the Quaternary Geology, Univ. of Turku.

Saarnisto, M. 1973. Contributions to the late-Quaternary history of the Lutto river valley, Finnish Lapland. Comm. Phys.-Mat. 43, 11-20.

Sissons, J. B. 1961. A subglacial drainage system by the Tinto Hills, Lanarkshire. Edinburgh Geol. Soc. Tr. 18 2, 175-193.

Stone, J. 1959. A description of glacial retreat features in Mid-Nithsdale. Scottish Geogr. Mag. 75, 3, 164-168.

Tanner, V. 1915. Studier öfver kvartärsystemet i Fennoskandias nordliga delar III. Bull. Comm. géol. Finlande 38.

Formation and Deformation of Glacial Deposits, Warren & Croot (eds) © 1994 Balkema, Rotterdam, ISBN 90 5410 096 6

Glaciotectonic structures along the southern Barents shelf margin

J. Sættem
IKU Petroleum Research, Norway

ABSTRACT: Glaciotectonic structures have been recognized on bathymetric, shallow seismic and core data from the outermost southwestern Barents shelf. A borehole located at 428 m water depth penetrates a 15 - 25 m thick interval of mid Cretaceous sedimentary rocks, resting on Pleistocene till. These rocks are interpreted to be part of buried glaciotectonic cupola-hills formed by a grounded ice sheet/ice stream which probably extended to the shelf break. An irregular bathymetry in a transverse trough in the southwesternmost corner of the Barents Sea is partly related to a glaciotectonic event which mainly affected glaciomarine/marine sediments. The glaciotectonics in this area are inferred from bathymetric and high resolution seismic data, partly by comparing the seismic signature with that of the cored structure farther north. The age of the described structures is not known in detail, but all were probably formed during the late Weichselian, and suggest several ice advances to the outermost shelf in the area in this period. Glaciotectonics is one of many phenomena which may be characterized by a chaotic internal reflection pattern on seismic data, and the cause for such a seismic signature can only be determined by coring or inferred by combining several lines of evidence.

1 INTRODUCTION

1.1 Offshore glaciotectonics

Glaciotectonic deformation as well as displacement of bedrock and drift masses has been known for a long time in formerly glaciated onshore areas (Aber 1988). Glaciotectonic structures may range in size up to 1000 km² in area and 200 m in structural relief, and are particularly common in the outer zones of glacially affected areas underlain by Mesozoic and Tertiary sedimentary rocks (Aber et al. 1989). Even so, observations of glaciotectonic forms and structures on the sedimentary rock areas of formerly glaciated continental shelves are fairly limited beyond the description of glaciotectonically influenced moraine ridges (e.g., Bugge et al. 1978; Solheim & Pfirman 1985; Boulton 1986; Josenhans & Fader 1989; Solheim 1991; Sættem 1990). Sættem (1990) recognized glaciotectonic hill-hole pairs (Bluemle 1970; Clayton & Moran 1974) about 1-1.7 km³ in size, at present 200 - 300 m water depths on the Norwegian continental shelf, and inferred the possible occurrence of buried glaciotectonic structures influencing both Quaternary and Tertiary sediments. Studies of coastal sections summarized by Aber et al. (1989) indicate that glaciotectonism may have a quite common relationship with marine based ice sheets, and evidence of glaciotectonics affecting pre-glacial sediments is recently also found in the southeastern Barents Sea (V.N. Gataullin & L.V. Polyak pers. comm., 1990).

The hitherto scarcity of literature on offshore glaciotectonics is probably due to very limited core data coverage and lack of criteria for recognizing such features in the marine environment by geophysical methods. The present paper intends to illustrate how glaciotectonic structures and features can be recognized by marine geological investigation methods using examples from the southwestern Barents Sea (Fig.1). Some stratigraphic/glacial geological implications of the described structures are dealt with briefly. The classification of the described glaciotectonic structures follows Aber et al.'s (1989) nomenclature.

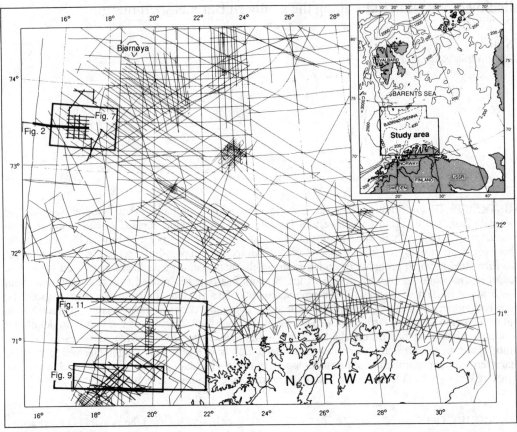

Figure 1 Study areas (Figs. 7,9,11) and the geoseismic profile along 73°30'N (Fig. 2) shown on a regional shallow seismic data base map for the southwestern Barents Sea.

1.2 Data and methods

The study is based on investigation of analog shallow seismic, bathymetric and borehole data. The core material was obtained from shallow boreholes drilled for the purpose of investigating preglacial bedrock, but short hammer samples are also recovered from the Quaternary sediment cover (Sættem et al. 1992). In borehole 7316/06-U-01 "difficult soil conditions" made it necessary to use a continuous rotary diamond coring technique also in the glacigenic sequence.

The regional shallow seismic data in the southern Barents Sea (Fig. 1) have been collected by various institutions and survey companies since 1970. The data quality varies a lot depending on seismic sources, recording parameters, recording conditions and geological conditions. Sparker is the most used seismic source, but mini airgun has also been used. The descriptions of the two areas in Figures 7 and 9 are mainly based on the analog mini airgun data, which tends to give better resolution and distinction of the seismic signature than the sparker. The resolution of the data (i.e. the ability to distinguish individual reflecting surfaces) is typically 5 - 15 s. Discrimination of point reflectors (i.e. erratics of metamorphic rocks or sedimentary rock floes) depends on seismic source frequency, lithologic contrasts, depth and noise level. The shallow seismic data have a dominating frequency in the 100 - 200 Hz range, which gives a Fresnel zone width of 100 - 140 m at 500 m depth. The reflection amplitude from bodies 25 - 35 m across will then be only 40% of that from bodies say ten times this width (cf., Badley 1985). A buried body less than 25 - 35 m across could probably still be recognized if the

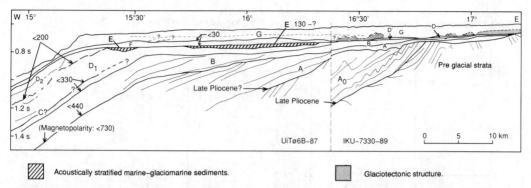

Figure 2 Prograding glacigenic sediments at the southwestern Barents Sea margin (after Sættem et al. 1991). The onset of glacigenic sedimentation is debated, but had according to Sættem et al. (1991, 1992) commenced in seismic unit A_0 time. The glaciotectonic structure within seismic unit G is described in the present paper.

noise level is low, or the acoustic impedance contrast to the surrounding sediments is high, but the potential for recognition of such bodies is often hampered by diffractions from sea bed irregularities like pockmarks and iceberg ploughmarks. Phenomena at a smaller scale may therefore remain undetected, particularly at some burial depth where seismic resolution is reduced and acoustic impedance contrasts may be reduced due to compaction.

The structure in Håkjerringdjupet (Fig. 9) is interpreted from a detailed bathymetric map (10 m contour interval) constructed by IKU based on soundings carried out by the Norwegian Hydrographic Service. In Figure 9 only 25 m contours are given.

2 SHALLOW GEOLOGY OF THE BARENTS SEA

The Barents Sea is an epicontinental sea in the northwestern corner of the Eurasian continent. Average present water depth is about 230 m (Elverhøi et al. 1989), but differential tectonic movements and glacial erosion and deposition have formed troughs and banks with water-depths ranging from less than 50 m close to Bjørnøya to more than 500 m in Bjørnøyrenna. A pronounced bathymetric slope along the coast follows the boundary between the Caledonian basement and the sedimentary rocks on the continental shelf.

A regional upper angular unconformity (URU) separates Tertiary and older preglacial sedimen-

tary rocks of varying dip from the overlying glacigenic sediments (e.g., Sundvor 1974; Bugge & Rokoengen 1976; Solheim & Kristoffersen 1984; Vorren et al. 1986). A large clastic wedge along the western shelf margin has been a major depocenter during the Late Cenozoic (e.g., Nansen 1904; Eldholm & Ewing 1971; Spencer et al. 1984; Eidvin & Riis 1989; Gabrielsen et al. 1990; Vorren et al. 1990a,b). This succession (Fig. 2) may comprise dominatly glacigenic sediments (Eidvin & Riis 1989; Sættem et al., 1991, 1992).

Only fragments are yet known of the glacial history of the Barents Sea, but a growing understanding is emerging from a number of studies of seismic and core data (Rokoengen et al. 1979a,b; Elverhøi & Solheim 1983; Solheim & Kristoffersen 1984; Vorren & Kristoffersen 1986; Vorren et al. 1989, 1990a; Hald et al. 1990; Sættem 1990; Sættem et al. 1991, 1992). Some of these studies suggest that grounded ice extended out to outer Bjørnøyrenna during the Weichselian, and the stratigraphy in Figure 2 suggests extensive Saalian and Weichselian deposition at the shelf margin.

3 BURIED CUPOLA-HILLS IN OUTER BJØRNØYRENNA, WESTERN BARENTS SEA

3.1 Borehole stratigraphy

Borehole 7316/06-U-01 (Figs. 3-5) is located at 428 m waterdepth and penetrates a 65 m thick

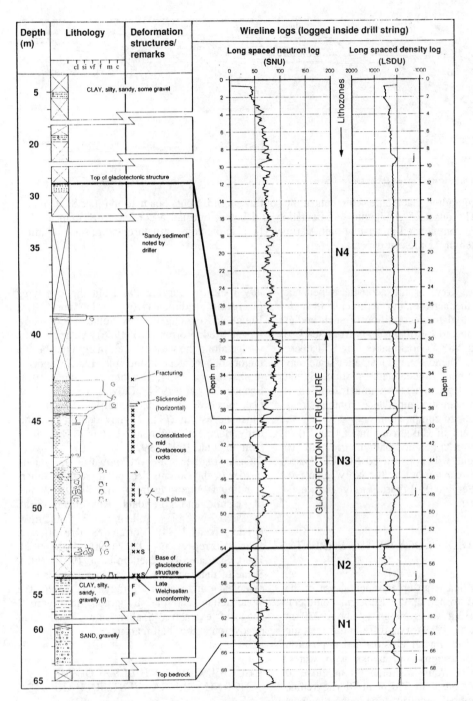

Figure 3 Stratigraphy of borehole 7316/06-U-01 showing mid Cretaceous rocks glaciotectonically emplaced in glacigenic sediments. Lithozone boundaries are according to Sættem et al. (1992), who assign a late Weichselian age to the unconformity at the base of the embedded Cretaceous rocks. The orientation and relative stratigraphic position within the glaciotectonic interval may be disordered due to glaciotectonic transportation. Wireline logs were run inside the drillstring, and drillpipe joints (j) are indicated on the neutron log.

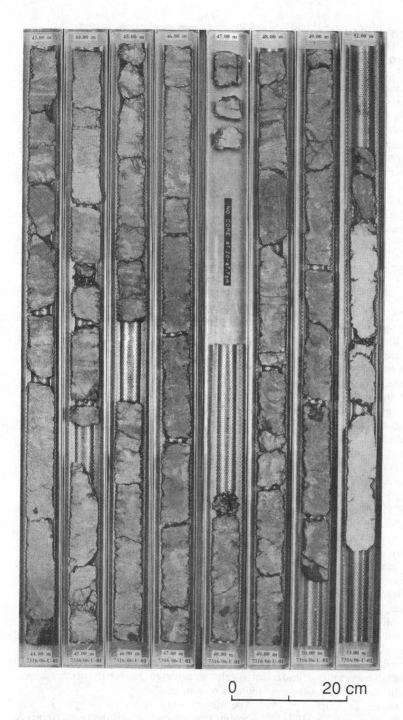

Borehole 7316/06–U–01

Allochthonous mid Cretaceous rocks embedded in Pleistocene glacigenic sediments

Figure 4 Core photo of allochthonous mid Cretaceous rocks in borehole 7316/06-U-01.

glacigenic succession on top of dipping preglacial sedimentary bedrock. The average core recovery is 21%. Embedded in the glacigenic sediments there is a 15 - 25 m interval of allochthonous preglacial sedimentary rocks. The lithology of this borehole is partly summarized from Sættem et al. (1992), together with a description of the allochthonous rocks. Lithozone N1 (Fig. 3) rests unconformably on sedimentary rocks, and comprises a gravelly sand with intervals of stiff - hard diamicton towards the top. The sand is interpreted as a possibly glacially reworked glaciofluvial sand, and it partly intercalates with the overlying diamicton interpreted as a till. The next lithozone (N2) is 5 m thick and yielded cores of a very stiff - hard muddy, partly fissile diamictic till. This till is sharply overlain by the interval of allochthonous sedimentary rocks at 54.1 m (Fig. 5). Between 54.1 m and 39.0 m only consolidated sedimentary rocks were recovered (Figs. 3,4). The rocks are partly intensely fractured, and although continuous coring was attempted the core recovery was only about 50 %. The drilling penetration rate in unrecovered intervals was partly very high, and most likely these intervals comprise loose (fractured?) rocks or/and unconsolidated sediments. Below 52.2 m the rocks consist of light grey sandy-silty claystone or marl. Clasts of glauconite and prisms of the bivalve Inoceramus are observed. This lithology is partly intensely deformed, and its upper boundary is probably tectonic. The interval between 52.2 m and 44.2 m contains a bioturbated, greyish silty shale with fragments and prisms of Inoceramus. This lithology apparently grades into the overlying sandstone. However, horizontal slickenside surfaces around 44.2 m show that this boundary may also be tectonic. A greenish to red-green mottled glauconite sandstone was cored between 39.0 m and 44.2 m. The interval is bioturbated, and only traces of primary bedding are preserved. Light grey-light olive grey sand was cored at 29.8-29.3 m and 39.0-38.9 m. The lower core may represent caving from the borehole, but the upper core which was recovered immediately after a marked drop in drilling fluid pressure, is probably in situ. Low fluid pressure and what the driller commented as "sandy conditions" continued down to 39 m, with a slight drop in penetration rate at 35 m. The neutron log (Fig. 3) suggests also a lithological change at about 30 m, although it cannot be used for direct interpretation of lithology or porosity, because

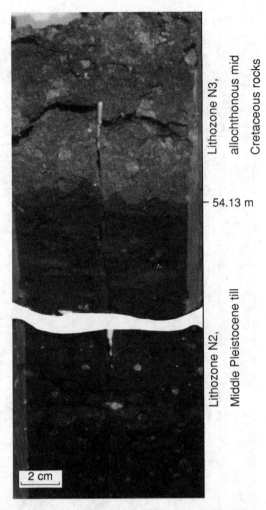

Figure 5 Tectonic boundary between glaciotectonically emplaced mid Cretaceous rocks and underlying Middle Pleistocene till. Fragments of the mid Cretaceous rocks are sheared into the till.

the wireline logs were run inside the drillstring without control on effects from hole geometry. On this basis the interval between 39 m and 29 m is also considered as a part of the allochthonous rock interval.

The deformation structures (Figs. 3-5) are probably not induced by drilling (e.g. Tibbits & Radford 1985), but rather result from brittle tectonic deformation during the displacement of these rocks. Within the recovered cores fracturing is more pronounced in the fine-grained lithologies (i.e. below 44.3 m) than in the sandstone interval, and both the grey siltstone and the marl are partly strongly brecciated. Both horizontal and vertical slickenside surfaces occur. A 1-2 cm wide, subvertical zone of soft fault gouge with angular fragments of intact rocks at about 49.5 m is bounded by far less fractured rocks. It may be older than the final displacement event, although the low consolidation of the fault gouge suggests that it has not been later subjected to high effective stress.

On top of the allocthonous rocks there is a 29 m thick interval of massive, homogeneous, dark grey muddy diamicton with 1 - 3% gravel. This diamicton has partly a horizontal fissility and a slight possible overconsolidation, and is interpreted by Sættem et al. (1992) as a till.

Biostratigraphy based on foraminifera, nannoplankton and dinoflagellates (unpublished IKU data) consistently assign a mid Cretaceous age to the rocks between 54.1 m and 39.0 m. The biostratigraphy of the sand cored at 29 m contains mixed assemblages including specimens indicative of glaciomarine influence, and the interval between 39 m and 29 m may comprise a mixture of Quaternary and pre-Quaternary sediments.

Borehole 7316/06-U-02, located less than 1.5 km to the north of 7316/06-U-01 (Fig. 6a), yielded cores of normally to slightly overconsolidated muddy, mostly massive diamictons from the glacigenic sediment interval. Some cores had a subhorizontal fissility, but sandy flames or a sand layer was found at one level. The cores from this borehole are interpreted as mostly finally glacially deposited (Sættem et al. 1992). Bedrock fragments in the cores from this borehole were only of gravel size, but the upper 1-2 metres of the preglacial bedrock were partly glaciotectonically deformed and contained small inclusions of Quaternary clay (Sættem et al. 1992).

3.2 Seismic mapping

The allochthonous rock interval in borehole 7316/06-U-01 correlates with the hummocky, acoustically chaotic unit in the lower part of seismic unit G in Figure 6. Northwards the hummocky seismic unit pinches out before borehole 7316/06-U-02, where no signs of bedrock rafts above gravel size were found. It is therefore assumed that the allochthonous rock interval in borehole 7316/06-U-01 corresponds to the hummocky seismic subunit. The geometry of the subunit is shown in Figure 7, and the seismic sections in Figures 6 and 8 illustrate acoustic expressions as well as a hummocky, and partly poorly defined surface.

The western part of the unit forms a number of amalgamated northwest-southeast trending ridges (Fig. 7), while an area in the east and a small outlier in the north cannot be conclusively connected with the remainder of the unit based on the existing seismic grid. Only in the north is the unit boundary partly clearly defined. Elsewhere the unit grades laterally into the lower part of the acoustically transparent unit G (Fig. 8), and local point reflections exist beyond the indicated unit boundaries. The unit is distinctly visible only on the analog airgun data, although it is crossed also by sparker profiles from several surveys (!). Velocity analyses on shallow digital seismic data (2.5 s record length) do not resolve any difference in sound velocity between the mapped unit and the surrounding glacigenic sediments.

The base of the mapped unit corresponds to the base of seismic unit G, which at least partly is developed as an erosional surface. Upwards the mapped unit is bounded by, and completely buried by the acoustically transparent unit G. Maximum thickness is about 55 ms (50 - 60 m) and, as mapped in Figure 7, it covers an area of about 600 km^2. The total volume cannot be confidently determined, but may be in the order of 10 km^3.

3.3 Interpretation

The allochthonous Cretaceous sedimentary rocks in borehole 7316/06-U-01 may in a deep water glacial/glaciomarine setting have been deposited by grounded ice, dropped from floating ice/ice bergs or have slid as olistoliths down a proglacial apron or the gentle northern flank of Bjørnøyrenna. There are, however, no indications that such aprons existed, and the slope of the underlying erosional surface is only 0.35°, which probably is too gentle to cause olistoliths to slide. The correlated seismic unit is interpreted as a constructional form protruding from an erosional surface of glacial origin, and the simplest explanation seems to be that this unit was deposited (or formed) in contact with grounded ice. Hence the allochthonous sedimen-

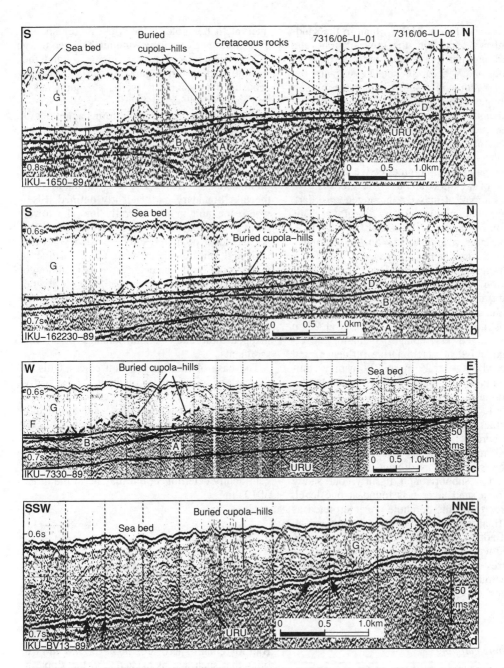

Figure 6 Examples of seismic expression (analog mini airgun profiles) of interpreted buried cupola-hills. Glaciotectonically emplaced mid Cretaceous rocks in borehole 7316/06-U-01 correlate with the humocky structure resting on the base of unit G (Fig. a), while in borehole 7316/06-U-02 only muddy glacigenic diamictons were found. The cupola- hills partly rest on preglacial bedrock, partly on glacigenic sediments. Arrows (Fig. d) point out possible velocity pull ups which may indicate discrete high velocity bodies within the hills. The unaffected reflections below the two northeasternmost indicated pull ups are probably diffractions. The seismostratigraphy of the glacigenic sequence is based on Sættem et al. (1991) (cf. Fig. 3 above).

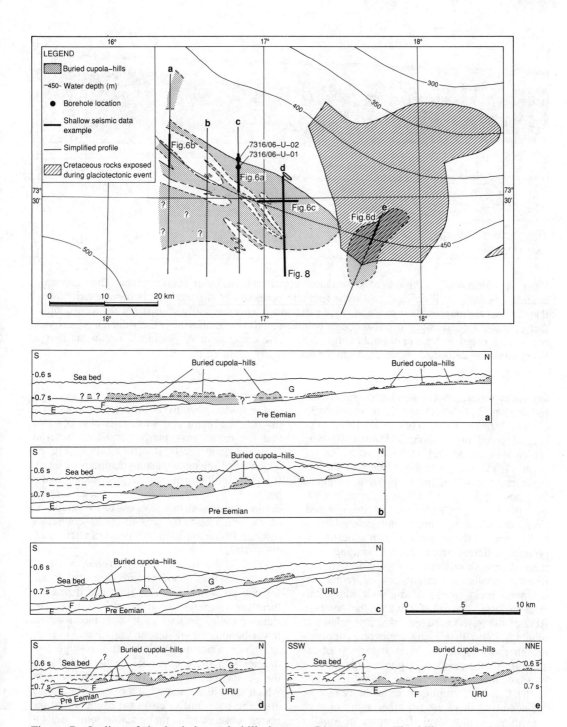

Figure 7 Outline of the buried cupola-hills in outer Bjørnøyrenna. The hills rest on the uncon-formity at the base of unit G. The exposure of Cretaceous rocks in the east is defined according to Sigmond (1992), but is herein bounded southwards by the coverage of glacigenic sediments older than the cupola-hills.

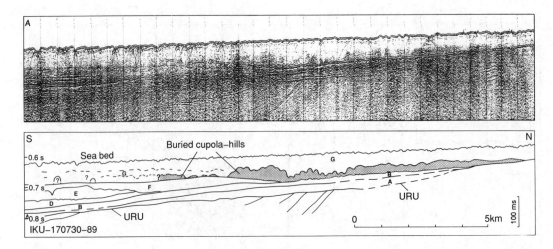

Figure 8 North south section across the buried cupola-hills in outer Bjørnøyrenna. The northern boundary is fairly well defined, while the southern boundary of the complex is transitional. Note that the hills are mappeable because they are surrounded by acoustically transparent sediments, and that a chaotic acoustic signature also occurs below the unit G unconformity. Embedded Cretaceous rocks were cored in the cupola-hills (Figs. 3-5), but a component of glacigenic sediments may partly comprise deformed sediments from the acoustically stratified unit E.

tary rocks were probably displaced, transported and deposited glaciotectonically as megablocks or floes (Stalker 1976; Aber et al. 1989). The interval cored in borehole 7316/06-U-01 may consist of one or several stacked floes.

The diffractions from the surface of the deposits mapped in Figure 7 appear to originate from positive forms and not from depressions as do relict iceberg ploughmarks on the sea bed. Hence the diffractions reflect upstanding hills or bodies, apparently of material with acoustic impedance different from the surrounding sediment. A smooth reflector in Figure 6b may represent a continuous megablock of rather undeformed rocks, about 2 km wide along this cross section. The relief towards the southern flank of the reflector suggest that it is related to lithology rather than being a gas brightspot, although gas may be localized at the top of the structure if it consists of a porous lithology. A possible smoothing by glacial erosion/moulding of the flat, high amplitude reflector cannot be excluded, but there are no other signs of such smoothing at this level. Examples of smaller discrete bodies (in the order of 100 m across) may occur above the arrows in Figure 6d. The proportion of megablocks or floes of any size in these deposits is unknown, but the stratigraphy of borehole 7316/06-U-01 and the seismic data

suggest that they may be a common constituent in the north, possibly becoming less important towards the south and west. It is also possible that the ridges may partly consist of material from the acoustically stratified unit E (Fig. 8) which has a seismic signature distinctly different from the unit covering the glaciotectonic deposits. Unit E was exposed to glacial erosion during the glaciotectonic event to the southeast of the glaciotectonic deposits, and comprises soft, underconsolidated sediments (Sættem et al. 1991, 1992).

There is no indication of erosional surfaces cutting into the glaciotectonic deposits, except for one possible case (Fig. 6b). The ridges are therefore largely constructional forms. The ridges parallel the axes of lows in the erosional lower boundaries of seismic units F and G (Fig. 7). Hence, a northwesterly ice flow is suggested during both these erosional events and during the formation of the glaciotectonic deposits which most likely started at the end of or closely after the base unit G erosional event. From the above observations the glaciotectonic unit is interpreted as being formed at the base of a grounded ice sheet or of the ice stream flowing out Bjørnøyrenna postulated by Grosswald (1980), here flowing in a northwesterly direction. In accordance with the nomenclature of

Aber et al. (1989), these deposits are therefore termed buried cupola-hills.

The source area for the displaced mid Cretaceous rocks must be located in an area where Cretaceous rocks were exposed to the ice sheet during the glaciotectonic event unless the transport occurred in several steps. Cretaceous rocks subcrop at base unit G level in the eastern part of Figure 7, and a depression in the bedrock surface (URU) in Figure 6d, may be a possible source area. However, megablocks are documented to have been transported glaciotectonically for more than 300 km (Jahn 1950; Ruszcynska-Szenajach 1976; Stalker 1976), and the source may be more distant.

4 A HILL-HOLE PAIR IN HÅKJERRING-DJUPET

4.1 Physiography

Håkjerringdjupet (Fig. 9) is a transverse trough in the southwesternmost corner of the Barents Sea. The continental shelf is here only 100 km wide. On the inner shelf the banks forming the flanks of the trough are partly bedrock surface (URU) forms, while towards the shelf break they are formed by glacigenic sediments. The boundary between glacigenic and preglacial sediments in this area is uncertain but could correspond to the unconformity indicated in Figure 9. Håkjerringdjupet is overdeepened both on the sea bed and at URU level, and is clearly a glacial erosional form. This was first pointed out by Holtedahl (1940), who also recognized the marked NE-SW trending escarpment in the central part of the trough, confirmed by later geophysical profiling to be located along a major bedrock fault, here separating Cretaceous sedimentary rocks downthrown to the west from older strata on the platform to the east (Gabrielsen et al. 1990). To the west of this escarpment both the sea bed and the URU are conspicuously irregular. The escarpment continues at URU level beneath the glacigenic sequence somewhat farther than expressed on the sea bed, and forms here the eastern flank of a buried NE-SW trending erosional valley.

4.2 Ice flow

Håkjerringdjupet has a rather streamlined bathymetry east of the escarpment (Fig. 9), while in the west it is characterized by overdeepened erosional depressions cut partly in sedimentary rocks, partly in glacigenic sediments. Drumlinoid sea bed forms casted in glacigenic sediments in the eastern part of the trough and flutes farther west indicate a trough-parallel flow in southern central Håkjerringdjupet (Fig. 9). A curved ridge on Sotbakken (Figs. 9,11) is tentatively interpreted by Vorren & Kristoffersen (1986) as part of Late Weichselian maximum moraines. This ridge, in accordance with Holtedahl's (1940) explanation, is here considered as being formed by a branch a Sørøysundet outlet glacier, and probably formed by ice push during a Late Weichselian readvance (Sættem 1990). The axes of drumlinoid sea bed forms in the central part of Håkjerringdjupet are oriented slightly to the south of west (i.e. radial to the curved ridges), suggesting an ice flow from Sotbakken. The direction of late ice flow in Håkjerringdjupet may therefore have varied between west-southwest and west-northwest.

4.3 Stratigraphy

The stratigraphy of the upper part of the glacigenic sequence in the area is shown in Figure (9b). Rokoengen et al. (1979a,b) have mapped glacigenic seismic units above what they call older drift (above unit T in Figure 9). In Håkjerringdjupet two of these units: Mulegga Drift east of the escarpment, and Nordvestsnaget Drift in the outer part of the trough, are both Upper Weichselian glacigenic sediments overrun by grounded ice (Rokoengen et al. (1979a,b). They consider the older drift as probably older than 18,000 BP.

A partly more detailed stratigraphy is proposed in Figure 9. Arguments for ages assigned to units are given in the discussion. The buried acoustically stratified unit (E') has a strong resemblence to a similar unit in outer Bjørnøyrenna (cf. Fig. 8) containing underconsolidated silty clay (Sættem et al. 1991, 1992), which according to unpublished work by the present author seems to be in a similar seismostratigraphic position. Till tongues (King & Fader 1986; King et al. 1987, 1991) occurring in the

a

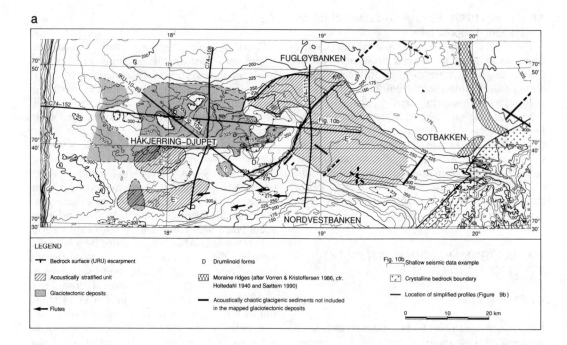

LEGEND

⊤ Bedrock surface (URU) escarpment

▨ Acoustically stratified unit

▦ Glaciotectonic deposits

◀ Flutes

D   Drumlinoid forms

▩   Moraine ridges (after Vorren & Kristoffersen 1986, cfr.
    Holtedahl 1940 and Sættem 1990)

—   Acoustically chaotic glacigenic sediments not included
    in the mapped glaciotectonic deposits

Fig. 10b Shallow seismic data example

⊡   Crystalline bedrock boundary

—   Location of simplified profiles (Figure 9b)

0   10   20 km

b

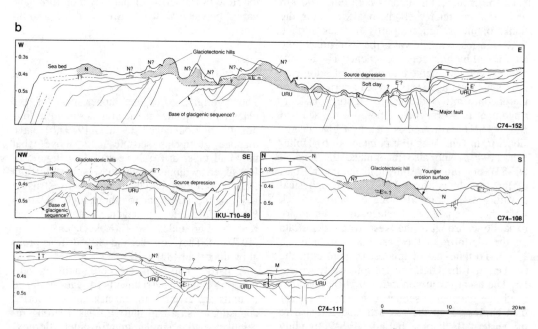

Figure 9 Outline of the buried acoustically stratified sediments and the inferred glaciotectonic hill-hole pair in Håkjerringdjupet. M and N denote the Mulegga and Nordvestsnaget Drifts (Rokoengen et al. 1979a). Units E' and T are defined herein. The glaciotectonic deposits probably largely comprise deformed acoustically stratified (marine - glaciomarine) deposits and younger till (units E' and T), and glaciotectonic decollement has mainly followed unit E'. Local erosion and deposition of the Nordvestsnaget Drift postdates the glaciotectonic event. Data examples are shown in Figure 10.

106

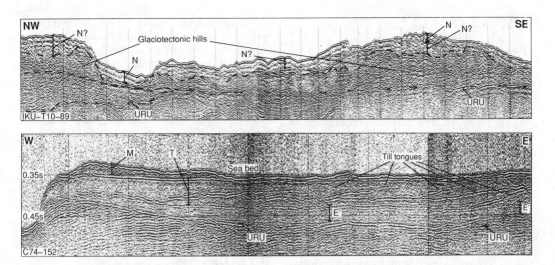

Figure 10 Examples of seismic signature of the glaciotectonic deposits in outer Håkjerringdjupet. Location of sections is shown in Figure 9a.

overlying unit, (T), in inner Håkjerringdjupet (Figs. 9b,10) suggest a glacial advance to the area along the surface of unit E'. Buried, acoustically stratified sediments also occur as erosional remnants in the outer part of Håkjerringdjupet (Fig. 10). These remnants are partly clearly cut by erosion, partly they grade into acoustically chaotic sediments. Late- and postglacial soft clay occupies depressions in the present sea bed (Rokoengen et al. 1977), in similar positions as the buried unit E'.

4.4 Glaciotectonics

In central Håkjerringdjupet units E' and T (Fig. 9) are cut by the marked sea bed escarpments. To the southwest of this escarpment the bathymetry is rather irregular due to partly buried, large, hummocky, acoustically chaotic deposits with numerous point reflectors. Several seismic lines show that the erosional remnants of acoustically stratified sediments become acoustically chaotic when cut by escarpments, and remnants of stratified sediments can be traced within the acoustically chaotic, hummocky deposits. The resemblence to the cupolahills in outer Bjørnøyrenna in morphology and acoustic signature suggests a similar origin. The association of hummocky hills and an up-ice depression may further suggest a genesis similar to the Steinbitryggen-Sopphola glaciotectonic

hill-hole pair located 50 km farther north (Sættem 1990). The irregular hills in outer Håkjerringdjupet are therefore interpreted to consist largelly of displaced sediments of units E' and T, which were removed from the depression farther east. Glaciotectonic decollement seems partly to have followed the lower part of the acoustically stratified unit which, like the similarly looking unit in outer Bjørnøyrenna, may comprise under-consolidated sediments. The sediments of this unit and any older glacigenic deposits have been removed from the depression close to the escarpment. Farther west the stratified sediments are crumbeled and mixed, and have been buried by sediments thrust and squeezed from the eastern depression.

The hills have a drumlinoid shape around 18°30'E, 70°40'N, and the shape of the glaciotectonic deposits were modified by subsequent erosion and the deposition of Nordvestsnaget Drift (Fig. 9b). The hummocky hills seem in the first place to have been less extensively developed in the southern part of Håkjerringdjupet than in the northern part, and later erosion and deposition have also been more extensive in the former area.

5 DISCUSSION

5.1 Recognition of glaciotectonics

The preceeding sections describe two examples of major glaciotectonic structures identified by borehole stratigraphy, internal acoustic signature and morphology. The buried cupola-hills in outer Bjørnøyrenna have no bathymetric expression, and no definite source area can be identified. The presence of allochthonous Cretaceous sedimentary rocks within these deposits is, together with the buried morphology, considered as conclusive evidence for at least a partly glaciotectonic origin of these deposits. The hill-hole pair in Håkjerringdjupet is interpreted as glaciotectonic in origin from the association of a source depression and hummocky acoustically chaotic down-ice depositional forms resembling the buried cupola hills in outer Bjørnøyrenna.

The escarpment across central Håkjerringdjupet has apparently formed the eastern boundary of the glaciotectonic decollement, although buried acoustically stratified (underconsolidated?) deposits also exist farther east (Fig. 9). Numerous pockmarks occur on the sea bed in the area. They probably indicate abundant fluid flow from the underlying preglacial bedrock (King & MacLean 1970; Hovland & Judd 1988). Such expulsion may be more intense in the faulted and folded area to the west of the escarpment (located at the margin of a basin) than on the platform to the east. The extensive bedrock erosion on the downthrown side of the escarpment may, from the characteristic irregular morphology and diffuse acoustic signature of the URU, also partly be ascribed to the above described or earlier glaciotectonic events, possibly here influenced by persistent fluid flow from the underlying basin (Sættem 1991). The URU morphology may, however, also reflect that the faulted and folded bedrock (Fig. 9b) has lateral variations in erodability, not necessarily related to glaciotectonic displacement.

The examples above illustrate that glaciotectonic deposits may be identified by reflection seismic profiling, but also that it is of outmost importance to use a seismic source with maximum resolution for the required penetration, which provides a distinct definition of seismic signature. Acoustic impedance contrasts associated with glaciotectonics will tend to be attenuated with increased burial depth because of equilibration of densities and sound velocities due to burial compaction. Also the resolution will decrease with depth because of the frequency filtering effect and the increase in Fresnel zone (Sheriff 1977). These effects reduce the possibility to recognize deeply buried glaciotectonic phenomena by seismic methods.

Seismic profiling gives also an opportunity to detect possible buried glaciotectonic structures which is unique to the marine environment. Glaciotectonics can, however, in general only be distinguished from many other phenomena with considerably uncertainty, unless the interpretation is supported by coring or by associations of forms/structures. Rather, glaciotectonics adds to a suite of possible explanations of seismic anomalies in glacigenic sediments (e.g. slump/slide masses, gas charged sediments, lithological variations, channelling).

A chaotic seismic signature may be related to glaciotectonism, but not necessarily to constructional glaciotectonic forms. Mounds penetrating into ice shelves are known to cause ice crevassing, and Vornberger & Whillans (1986) and Shabtaie et al. (1987) mapped the boundaries of ice streams feeding the Ross Shelf as up to 20 km wide shear zones with intense ice crevassing. Crevasse zones are, according to Vornberger & Whillans (1986), associated with transverse variation in basal shear stress, and it is therefore likely that they are expressed also in the glacier base geometry. A complex stress field at the ice-substratum interface will cause complex variations in basal thermal conditions (cf. Røthlisberger & Iken 1981) and, if the substratum is deformable, an irregular glacier base geometry. Such stress and thermal conditions may result in chaotic spatial variations in consolidation (Sættem 1990), and hence acoustic impedance, of the substratum. Sediments, which elsewhere are acoustically transparent or stratified, may under such conditions become acoustically chaotic. This may have been the case in the transition between the buried cupola-hills and overlying diamicton in Bjørnøyrenna. From seismic expression and the borehole data, the bulk of the two glaciotectonic structures described herein are still interpreted as true constructional morphological forms.

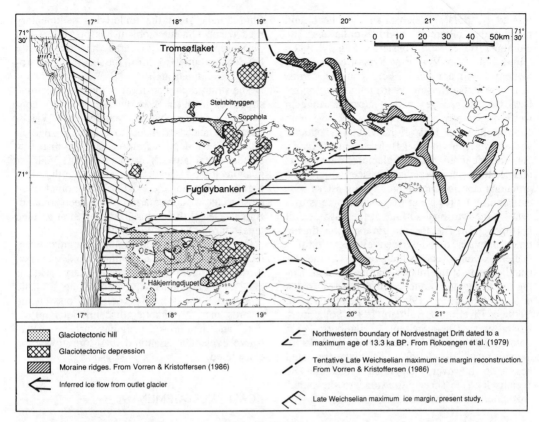

Figure 11   Glaciotectonic sea bed forms in the southwesternmost Barents Sea (modified after Sættem 1990). The maximum extent of the late Weichselian ice sheet is herein suggested to have followed the shelf break. Two other ice margin positions probably represent readvances during a general recessional phase. Arrows indicate inferred ice stream branches fed by Holtedahl's (1940) "Sørøysundet outlet glacier".

5.2 Age of the glaciotectonics and late Weich-selian ice sheet extent

The buried cupola-hills in outer Bjørnøyrenna mainly rest on the base unit G, erosional uncon-formity (Fig. 7), and probably formed at the end of or shortly after the erosional event. Both this erosional surface and the deeper base unit F un-conformity are dated to be of late Weichselian age (Sættem et al. 1991, 1992), and probably both formed during early expansion(s) of the late Weichselian ice sheet. In the southern Bar-ents Sea the presence of this ice sheet is con-fined to between approx. 27-30,000 years BP (Andreassen et al. 1984, 1985; Hald et al. 1990) and 13,290 years BP (Vorren & Kristoffersen 1986).

Erosional remnants of units E' and T in the northern flank of Håkjerringdjupet show that the glaciotectonic structure here is younger than both these units. The even younger Upper Weichselian Nordvestsnaget and Mulegga Drifts of Rokoengen et al. (1979a) superpose the top unit T unconformity (Fig. 9b), which extends northwards across Fugløybanken and here is cut by the glaciotectonic event forming the Steinbit-ryggen-Sopphola hill-hole pair (Sættem 1990; Fig. 11). Sættem (1990) correlated this uncon-formity, which probably is diachronous, with the base of the upper Weichselian unit b4 of Hald et al. (1990). Later investigations (Sættem et al. 1992) support this age.

From the stratigraphic relations it seems that both the glaciotectonic hill-hole pairs and the isolated depressions shown in Figure 11 were formed by a late Weichselian ice sheet, and possibly they all are of roughly similar age. The moraine ridges on eastern Fugløybanken (Holtedahl 1940; Vorren & Kristoffersen 1986; Sættem 1990) are most likely younger forms. Vorren & Kristoffersen (1986) tentatively correlated the latter ridges with a moraine ridge in western Bjørnøyrenna (Elverhøi & Solheim 1983; Solheim & Kristoffersen 1984). Together with the structure in outer Bjørnøyrenna (Fig. 7), these define three different glaciotectonic events related to a late Weichselian ice which was grounded near the shelf edge in the western Barents Sea. A fourth event of grounded ice is evident by the erosional surface at the base of unit F (Figs. 2,7). The ice margin position during times between the described events is still a matter of controversy which relates to the debated genesis of the muddy diamictons dominating the glacigenic sequence in the area (Vorren et al. 1978, 1989, 1990a; Rokoengen et al. 1979a,b; Elverhøi & Solheim 1983; Solheim & Kristoffersen 1984; Sættem & Hamborg 1987; Solheim et al. 1988; Elverhøi et al. 1989, 1990; Sættem et al. 1991, 1992). The data presented herein do, however, clearly show that the late Weichselian ice sheet did extend to the outermost continental shelf.

6 CONCLUSIONS

The present study illustrates how glaciotectonic features can be detected by marine geological investigation methods.

A borehole at more than 400 m water depth in outer Bjørnøyrenna penetrates a 15 - 25 m interval of mid Cretaceous sedimentary rocks embedded in Upper Pleistocene glacigenic sediments. The interval is correlated with seismically mapped buried cupola-hills, forming several up to about 50 m high NW-SE trending ridges, within an area of about 600 km^2, which were formed during the late Weichselian.

Irregular hills less than 100 m high, covering an area of about 500 km^2 in the outer part of the bathymetric trough Håkjerringdjupet, are related both to glaciotectonism and later glacial erosion. The hills probably mainly comprise deformed glaciomarine sediments and till displaced from a neighbouring depression to the east. Glaciotectonic decollement has apparently followed a buried acoustically stratified (underconsolidated?) unit, and has possibly been enhanced by

fluid escape from the underlying sedimentary rock basin. Glaciotectonism has possibly taken place repeatedly in this area, and may partly have contributed to form the bedrock escarpment across the trough.

The stratigraphic position of the glaciotectonic structures in outer Bjørnøyrenna provides conclusive evidence that a grounded late Weichselian ice sheet extended to the outer shelf in this area at a depth of more than 500 m below present water level. Together with the younger glaciotectonic structures farther south, their presence confirms the existence of grounded ice close to the western margin of the deep southern Barents Sea at three, possibly four, stratigraphic levels during the late Weichselian.

Glaciotectonics are difficult to recognize when the resulting structures neither make sea bed forms nor affect preglacial sedimentary rocks, or glacigenic sediments with contrasting acoustic properties. The typical expression on seismic records may be chaotic, discontinuous reflections, and hummocky, irregular surfaces. Alternative explanations should therefore always be considered.

ACKNOWLEDGEMENTS

The present paper was written as part of a Dr. Ing. thesis work at the Department of Geology and Mineral Resources Engineering, Norwegian Institute of Technology, University of Trondheim, financed by Statoil and IKU Petroleum Research. The work has benefitted from discussions with collegues at IKU. A. Mørk helped with the sedimentologic description of the allochthonous sedimentary rock rafts in borehole 7316/06-U-01, and the biostratigraphy of these rocks was investigated by R.M. Goll, M. Smelror and J.G. Verdenius. The manuscript has been critically read by K. Rokoengen. The English text was corrected by Stephen Lippard and the figures were drawn by Berit Fossum. The collection of the IKU Barents Sea Mapping Program data (stratigraphic borehole and shallow seismic data collected since 1984) has been financed by a number of oil companies and the Norwegian Petroleum Directorate (NPD). In particular I wish to acknowledge the permission to use data from an ongoing project granted by A/S Norske Shell, BP Norway, Conoco Norway Inc., Elf Aquitaine Norge A.S., Fina Exploration Norway, Mobil Exploration Norway Inc., Norsk

Hydro a.s, Phillips Petroleum Co. Norway, Saga Petroleum a.s and Statoil. To all the above institutions, companies and persons I direct my sincere thanks.

REFERENCES

Aber, J.S. 1988: Bibliography on glaciotectonic references. In D.E. Croot (ed.): *Glaciotectonics: Forms and Processes*, 195-210. A.A. Balkema, Rotterdam.

Aber, J.S., Croot, D.G. & Fenton, M.M. 1989: *Glaciotectonic landforms and structures*. Klüwer academic publishers, Dordrecht. 201 pp.

Andreassen, K., Vorren, T.O. & Johansen, K.B. 1984: Pre sen Weichsel glasimarine sedimenter på Arnøy, Nord Norge. *Meddelande från Stockholms Universitets Geologiska Institusjon* 255, 9, (abstract).

Andreassen, K., Vorren, T.O. & Johansen, K.B. 1985: Pre-Late Weichselian glacimarine sediments at Arnøy, North Norway. *Geologiska føreningen i Stockholm Førhandlingar 107*: 63-70.

Badley, M.E. 1985: Practical seismic interpretation. *International Human Resources Development* Corporation Publishers, Boston, 266pp.

Bluemle, J.P. 1970: Anomalous hills and associated depressions in Central North Dakota. *Geological Society America, abstracts with Programs* 2: 325-326.

Boulton, G.S. 1986: Push-moraines and glacier-contact fans in marine and terrestrial environments. *Sedimentology* 33: 677-698.

Bugge, T. & Rokoengen, K. 1976: Geologisk kartlegging av de øvre lag på kontinentalsokkelen utenfor Troms. *Continental Shelf Institute, publication 85*, 51 pp. (English summary.)

Bugge, T., Lien, R. & Rokoengen, K. 1978: Kartlegging av løsmassene på kontinentalsokkelen utenfor Møre og Trøndelag: seismisk profilering. (Quaternary deposits off Møre and Trøndelag, Norway: seismic profiling). *Continental Shelf Institute, publication 99*, 55 pp.

Clayton, L. & Moran, S.R. 1974: A glacial process-form model. In D.R. Coates (ed.): Glacial geomorphology. SUNY-Binghampton. Published in *Geomorphology*, Binghampton, New York: 89-119.

Eidvin, T. & Riis, F. 1989: Nye dateringer av de tre vestligste borehullene i Barentshavet. Resultater og konsekvenser for den Tertiære hevingen. NPD-Contribution 27; 44 pp, 9 figures, 5 plates.

Eldholm, O. & Ewing, J. 1971: Marine geophysical survey in the southwestern Barents Sea. *Journal of Geophysical Research 76*: 3832-3841.

Elverhøi, A. & Solheim, A. 1983: The Barents Sea ice sheet, a sedimentological discussion. *Polar Research 1* n.s.: 23-42.

Elverhøi, A., Pfirman, S.l., Solheim, A. & Larssen, B.B. 1989: Glaciomarine sedimentation in epicontinental seas exemplified by the northern Barents Sea. *Marine Geology 85*: 225-250.

Elverhøi, A., Nyland-Berg, M., Russwurm, L. & Solheim, A. 1990: Late Weichselian ice recession in the central Barents Sea. In U. Bleil and J. Tiede (eds.): *Geologic history of the Polar Oceans: Arctic versus Antarctic. NATO ASI Series C: Mathematical and Physical Sciences-volume 308*. Klüver Acad. Publ., Dordrecht: 289-307.

Gabrielsen, R.H., Færseth, R.B., Jensen, L.N., Kalheim, J.E. & Riis, F. 1990: Structural elements of the Norwegian Continental Shelf. Part 1 The Barents Sea Region. *NPD Bull 6*.

Grosswald, M.G. 1980: Late Weichselian ice-sheet of Northern Eurasia. *Quaternary Research 13*: 1-32.

Hald, M., Sættem, J. & Nesse, E. 1990: Middle and Late Weichselian stratigraphy in shallow drillings from the southwestern Barents Sea: foraminiferal, amino-acid and radiocarbon evidence. *Norsk Geologisk Tidsskrift 70*: 241-257.

Holtedahl, O. 1940: The submarine relief off the Norwegian coast. *Det Norske Videnskaps-Akademi*, Oslo, 43pp.

Hovland, M. & Judd, A.G. 1988: *Seabed pockmarks and seepages. Impact on geology, biology and the Marine environment*. Graham and Trotman Limited, London, 293 pp.

Jahn, A. 1950: Nowe dane o polozeniu kry jurajskiej w Lukowie (New facts concerning the ice transported blocks of the Jurassic at Lukow). *Annales Societatis Geologorum Poloniade 19*: 372-385, Krakow.

Josenhans, H.W. & Fader, G.B.J. 1989: A comparison of models of glacial sedimentation along the eastern Canadian margin. *Marine Geology 85*: 273-300.

King, L.H. & MacLean, B. 1970: Pockmarks on the Scotian shelf. *Geological society of America Bulletin 81*: 3141-3148.

King, L.H. & Fader, G.B. 1986: Wisconsinan glaciation of the continental shelf - southeast Atlantic Canada. *Geological Survey of Canada, Bulletin 363*, 72 pp.

King, L.H., Rokoengen, K. & Gunleiksrud, T. 1987: Quaternary seismostratigraphy of the Mid Norwegian Shelf, 65gr - 67gr 30min N. A till tongue stratigraphy. *Continental Shelf Institute, publication 114*, 58 pp.

King, L.H., Rokoengen, K., Fader, G.B., & Gunleiksrud, T. 1991: Till-tongue stratigraphy. *Geological Society of America Bulletin 103*: 637-659.

Nansen, F. 1904: The bathimetrical features of the North Polar seas, with a discussion of the continental shelves and previous oscillations of shoreline. In *"The Norwegian North Polar Expedition 1893 - 1896. Scientific Results."* Vol. 4. Christiania.

Rokoengen, K., Bell, G., Bugge, T., Dekko, T., Gunleiksrud, T., Lien, R., Løfaldli, M. & Vigran, J.O. 1977: Prøvetaking av fjellgrunn og løsmasser utenfor deler av Nord-Norge i 1976 (Sampling of bedrock and Quaternary deposits offshore North-Norway in 1976). *Continental Shelf Institute publication 91*, 67 pp. (In Norwegian, with English summary).

Rokoengen, K., Bugge, T. & Løfaldli, M. 1979a: Quaternary geology and deglaciation of the continental shelf off Troms, North Norway. *Boreas 8*: 217-227.

Rokoengen, K., Bugge, T., Dekko, T., Gunleiksrud, T., Lien, R. & Løfaldli, M. 1979b: Shallow geology of the continental shelf off North Norway. *Port and ocean engineering under arctic conditions*, Trondheim, Norway 1979, 17 pp.

Ruszczynska-Szenajch, H. 1976: Glaciotectonic depressions and glacial rafts in mid-eastern Poland. In Rozycki, S. Z. (ed.), *Pleistocene of Poland*. Studia Geologica Polonica I, Warsaw: 87-106.

Røthlisberger, H & Iken, A. 1981: Plucking as an effect of water-pressure variations at the glacier bed. *Annals of Glaciology 2*: 57-62.

Shabtaie, S,. Whillans, I.M. & Bentley, C.R. 1987: The morphology of ice streams A, B and C, West Antarctica, and their environs. *Journal of Geophysical Research 92*, No B9: 8865-8883.

Sheriff, R.E. 1977: Resolution of seismic reflections and geologic detail derivable from them. In C.E. Payton (Ed.): *Seismic stratigraphy - applications to hydrocarbon exploration*. AAPG memoir 26: 3-14.

Sigmond, E.M.O., 1992: Bedrock map of Norway and adjacent ocean areas. Scale 1:3 million. *Geological Survey of Norway*.

Solheim, A. 1991: The depositional environment of sub-polar tidal glaciers: A case study of the morphology, sedimentation and sediment properties in a surge affected marine basin outside Norsdaustlandet, northern Barents Sea. *Norsk Polarinstitutt Skrifter 194*, 97pp.

Solheim, A. & Kristoffersen, Y. 1984: Distributions of sediments above bedrock and glacial history in the western Barents Sea. *Norsk Polarinstitutt Skrifter 179B*.

Solheim, A. & Pfirman, S. L. 1985: Sea-floor morphology outside a grounded, surging glacier, Bråsvellbreen, Svalbard. *Marine Geology 65*: 127-143.

Solheim, A., Milliman, J.D. & Elverhøi, A. 1988: Sediment distribution and sea-floor morphology of Storbanken: implications for the glacial history of the northern Barents Sea. *Canadian Journal of Earth Sciences 25*, No. 4: 547-556.

Spencer, A.M., Home, P.C. & Berglund, L.T. 1984: Tertiary structural development of the western Barents Shelf: Troms to Svalbard. In A.M. Spencer et al. (eds.): *Petroleum geology of the North European Margin*. Norwegian Petroleum Society. Graham & Trotman, London: 199-209.

Stalker, A.Mac.S. 1976: Megablocks, or the enormous erratics of the Albertan Prairies. *Geological Survey of Canada, Paper No. 76-1C*: 185-188.

Sundvor, E. 1974: Seismic refraction and reflection measurement in the southern Barents Sea. *Marine Geology 16*: 255-273.

Sættem, J. 1990: Glaciotectonic forms and structures on the Norwegian continental shelf: observations, processes and implications. *Norsk Geologisk Tidsskrift 70*: 81-94.

Sættem, J. 1991: Glaciotectonism - an important process in Late Cenozoic erosion in the southwestern Barents Sea. In Sættem, J.: *Glaciotectonics and glacial geology of the southwestern Barents Sea*. Dr. ing. Thesis, Norwegian Institute of Technology, University of Trondheim: 75-115. Unpublished.

Sættem, J. & Hamborg, M. 1987: The geological implications of the upper seismic unit, south-eastern Barents Sea. *Polar Research 5*. n.s.: 299-301.

Sættem, J., Poole, D.A.R, Sejrup, H.P. & Ellingsen, K.L. 1991: Glacial geology of outer Bjørnøyrenna, western Barents Sea: preliminary results. *Norsk Geologisk Tidsskrift 71*: 173-177.

Sættem, J., Poole, D.A.R, Sejrup, H.P. & Ellingsen, K.L. 1992: Glacial geology of outer Bjørnøyrenna, southwestern Barents Sea. *Marine Geology 103*: 15 - 51.

Tibbits, G.A. & Radford, S.R. 1985: New technology for the recovery of representative cores from uncemented sand formations. *Proceedings 60th Annual Technical conference and Exhibition of the Society of Petroleum Engineers* (SPE 14297), Las Vegas NV September 1985, 12 pp.

Vornberger, P.L. & Whillans, I.M. 1986: Surface features of ice stream B, Marie Byrd Land, West Antarctica. *Annals of Glaciology 8*: 168-170.

Vorren, T.O., Strass, I.F. & Lind-Hansen, O.W. 1978: Late Quaternary Sediments and Stratigraphy on the Continental Shelf off Troms and west Finnmark, Northern Norway. *Quaternary Research 10*: 340-365.

Vorren, T.O. & Kristoffersen, Y. 1986: Late Quaternary glaciation in the south-western Barents Sea. *Boreas 15*: 51-59.

Vorren, T.O., Kristoffersen, Y. & Andreassen, K. 1986: Geology of the inner shelf west of North Cape, Norway. *Norsk Geologisk Tidsskrift 66*: 99-105.

Vorren, T.O., Lebesbye, E., Andreassen, K. & Larsen, K.B. 1989: Glacigenic sediments on a passive continental margin exemplified by the Barents Sea. In R.D. Powell and A. Elverhøi (eds.): Modern Glacimarine Environments: Glacial and Marine Controls of Modern Lithofacies and Biofacies. *Marine Geology 85*: 251-272.

Vorren T.O., Lebesbye, E. & Larsen, K.B. 1990a: Geometry and genesis of the glacigenic sediments in the southern Barents Sea. In J.A. Dowdeswell & J.D Scourse (eds.): *Glacimarine environments, processes and sediments. Geological Society of London, Special Publication 53*: 309-328.

Vorren, T.O., Richardsen, G., Knutsen, S.M. & Henriksen, E. 1990b: The western Barents Sea during the Cenozoic. In Bleil, U. and Thiede, J. (eds.): *Geologic history of the Polar Oceans: Arctic versus Antarctic.* NATO ASI Series C: Mathematical and Physical Sciences-volume 308. Klüver Acad. Publ., Dordrecht: 95-118.

Observations on drumlinized till in the Rhine glacier area (South German Alpine foreland)

Dietrich Ellwanger
Geologisches Landesamt Baden-Württemberg, Freiburg i. Br., Germany

ABSTRACT: This paper is a case study of longitudinal and cross sections of two drumlins in different stratigraphic and morphological, but similar lithological, settings. Common characteristics are: a glacial erosion surface of the underlying gravels, leeward-dipping till layers in depressions of the gravel surface, and horizontal drag of whole blocks of gravels within or beneath the till. Clastic intrusions of fine material into gravels and sand help to identify conditions in the subglacial environment and the pattern of deformation.

1. INTRODUCTION

Lithofacies and glaciotectonic deformation are subject of studies in formerly glaciated areas of the Pleistocene Rhine glacier. One of the aims is to compare the deposits of an older glacial complex of Riss age with those of the youngest (Würm). In this paper, investigations mainly focus on a site in the Riss moraine area, near the village of Weihwang, north of Pfullendorf (site 1, Fig. 1). The site is situated on the slopes of a Southern tributary valley of the River Danube, about half way between the main terminal moraines of Riss and Würm age. Like other tributaries, this valley is cut into an elevated plain capped by till. Moraine hills are generally of low amplitude, and whilst many are elongated in the direction of ice flow (and are considered to be drumlins), other ridges are oriented oblique or transverse to former ice flow directions. At the Weihwang site, drumlinized till overlies fluvial gravels of a buried river channel of the same glacial complex, i.e. of Riss age.

As, primarily, drumlin is a term describing a landform, the morphology of the whole area is briefly discussed. Then the sequence of gravels and till of the Weihwang site is considered: first in stratigraphic profile, then lateral variations. This sequence is finally compared with a similar sequence at the Hirschbrunnen drumlin site near Markelfingen which is increasingly becoming one of the key study sites with regard to drumlin formation in the Rhine glacier area (e.g. Schreiner 1973, 1974, Ellwanger 1990).

2. MORPHOLOGY OF THE RISS MORAINE AREA

The area of Riss moraines (classically described by Penck & Brückner 1901/09, and modern stratigraphic definition by Schreiner & Haag 1982, Schreiner 1989) is situated between the terminal moraines of the Würm glaciation and the Jurassic highlands of the Swabian Alb (Fig. 1). It is an almost featureless morainic plain, the till sitting either on soft Molasse bedrock or on Pleistocene gravels which fill former valleys incised into the bedrock. The plain's surface rises up to 50 m above today's valleys and dips slightly north (down ice) towards the Danube. Some of the outermost, more or less isolated moraines of Riss age are situated north of the Danube on the slopes of the Swabian Alb.

The pattern of today's valleys primarily follows depressions between moraine hills, parallel to the former direction of ice movement, so that much of the original morphological setting of the hills is still recognizable in spite of the large amount of erosion that happened since. Fig. 2 shows the orientation of hills and ridges.

In some areas, transverse ridges dominate. As they often consist of deformed or undeformed Quaternary deposits, they used to be interpreted as "end-moraines" (e.g.

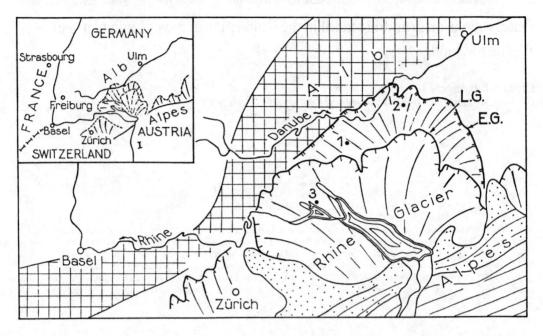

Fig. 1: Rhine glacier, location of study sites and areas. 1 = Weihwang drumlin, 2 = Dürmentingen drumlin field, 3 = Markelfingen drumlins. LG = maximum extent of ice during last glaciation, EG = maximum extent of ice during earlier glaciations. The area of Riss moraines lies between LG and EG.

Werner 1975). Some ridges of similar morphology, however, consist mainly of Molasse bedrock and therefore may be of erosional, probably subglacial origin.

There are two areas where a majority of moraine hills are elongated in the direction of ice flow: (1) the area north of Pfullendorf, and especially (2) the area around Dürmentingen (areas 1 + 2, Fig. 1). Here, the term "drumlin field" seems morphologically appropriate.

In the area around Dürmentingen, the morphology has only been slightly altered by subsequent erosion. Elongation of most hills and interdrumlin hollows in the direction of ice flow is obvious (Fig. 2). There is spatial variation in drumlin form, apparently related to separate ice lobes which can be reconstructed from the planform of the end-moraines north of Dürmentingen. The length/width ratios of drumlins decreases towards the centre of these reconstructed lobes, away from the interlobate junctions.

In general within the Rhine glacier area, Riss drumlins appear to be longer than Würm aged drumlins: well preserved Riss drumlins show a mean length/width ratio of 3/1,

whereas Würm drumlins have a value of 2/1. Therefore, another question to be investigated is whether this difference arises from depositional processes involved in the formation of Riss and Würm aged drumlins, or whether the morphological differences merely indicate the difference in geographic positions, within (Würm) and outside (Riss) the Lake Constance glacial basin.

Both East and West of the Dürmentingen drumlin field, further drumlin-like landforms can be identified, but these are mixed with transverse and oblique ridges. It is only to the north of Pfullendorf, that topographic orientation in the direction of former ice movements is evident. However in this area postglacial erosional processes have been very active, and it may be unwise to call this a "drumlin field", although individual landforms are definitely drumlins.

A drumlin to the east of Weihwang (in a gravel pit north of Pfullendorf) has been partly eroded, and a good section exposed river channel deposits and their drumlinized till cover.

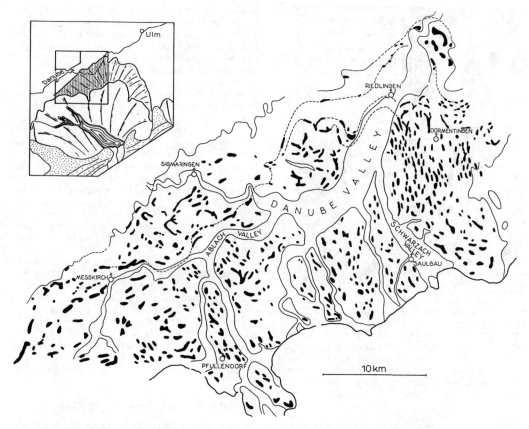

Fig. 2: Orientation of landforms in the area of Riss moraines. Typical drumlin fields are only near Dürmentingen and near Pfullendorf.

3 THE WEIHWANG DRUMLIN

3.1 Stratigraphic Sequence

The stratigraphic sequence of the deposits at Weihwang is shown in Fig. 4. The lower part of the section is dominated by gravels, the upper by till. The gravel part fills up a distinct river channel of a former valley whilst the till dominated sequence consists of interstratified layers of diamicton, gravel and sand.

The lower element can be further subdivided: The base comprises massive, matrix-rich, coarse-grained gravels, above which are stratified, better sorted medium-grained gravels, deposited in lense shaped minor channels. Some pebbles consist of cemented gravel: they are derived from deposits of still older glaciations (Cover Gravel Complex, Fig. 3).

A transition zone of cross bedded layers follows. Deposits are dominated by sand; gravels are fine to medium grained. Small pieces of reworked wood, layers of silt, and blocks of reworked interglacial soil material (the latter from a corresponding horizon in a nearby section), also occur. This transition zone not only indicates increasing distance from the glacier but also a temperate, nonglacial phase between deposition of the gravels beneath and above.

The gravel sequence continues with upward-coarsening beds of gravels (proximal outwash of advancing glacier).

The upper parts of the gravel sequence are cemented in varying degrees by a carbonate matrix. At the very top, a lithified layer of 10-40 cm consists mainly of sand and fine gravel, with some coarser pebbles; sometimes pebble nests also occur. Some of it may be derived from gravels immediately beneath, but appears structurally different.

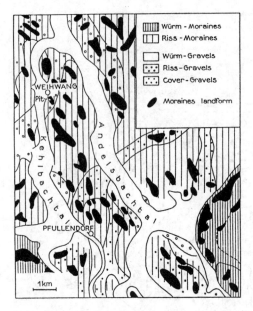

Fig. 3: Geological situation of the Weihwang site. Riss aged gravels within former
River channel ("buried valley") at the location itself, "Cover Gravels" (deposits of
earlier glaciations) beneath till further
South. In areas where neither is indicated,
soft Molasse bedrock lies immediately
beneath the till cover.

As previously indicated, the gravel sequence was deposited into a distinct river
channel cut into Molasse bedrock. This
channel can be followed right to the
Danube. Similar sequences (which can often
be correlated on the basis of lithology)
are found in several exposures in the area,
cutting older river channels of Riss age.

In contrast, the facies overlying the
river gravels at Weihwang, are "unique",
i.e. correlation of distinct layers from
one exposure to the next is not possible.
These glacial and fluvial deposits change
rapidly, both vertically and laterally.
Glaciotectonic disturbances also occur. The
till-dominated upper part of Fig. 4 therefore represents only the sequence in the
area immediately surrounding the site.

Two diamictons are differentiated on the
basis of their colour: light brown occurs
in lower horizons, light yellow further up.

Within lower parts of the light brown
diamicton, blocks and layers of cemented
gravels are intercalated, and most probably
represent material reworked from beneath.
Further up, interbedded layers of sand and

gravels (with pebbles up to 30 cm) are
found. Further light brown diamicton layers
follow, capped by lenses and layers of fine
material (sand and silt), interbedded with
coarse sands and gravels which are occasionally cross-bedded.

The latter form a transition zone towards
a coarse horizon of usually strongly cemented gravels. Large pebbles (up to 25 cm)
are often identified as local material
(clasts of Molasse bedrock, of Tertiary
limestone, and of reworked Cover Gravel
clasts). Some sedimentary structures are,
in spite of cementation, still recognizable
(indistinct bedding planes and a few
imbricated pebbles). These gravelly intercalations are only local accumulations
(subglacial or ice marginal) within the
till cover sequence. At the top, again
10-50 cm of lithified gravelly material
occurs.

The yellow diamicton appears layered by
interbedded sands and gravels, and its matrix content decreases upwards (Dm - Dc).
This is considered to be a till sequence
developed by down-melting glacier (e.g.
Grube 1980). The upper layers are strongly
weathered (decalcified); they consist of
diamicton and sand mixed up with loess. A
red-brown fossil soil is clearly developed
(Eemian?).

3.2 Lateral variations of the Till

The range of lateral variations in the
till-dominated part of the sequence is
shown in Fig. 5. They have been produced by
variations in depositional environments and
by glaciotectonic deformation and dislocation. With regard to the drumlin landform,
the section is longitudinal in a north-
south direction and cross-sectional at the
lee in a northeast-southwest direction.

3.2.1 Deformation of Gravel-Intercalations

The way the gravel intercalations were deformed (i.e. reacted to the applied glacial
stress), changes from stoss to lee (Fig.
5). Near the stoss end strongly folded
layers occur, whereas further towards lee,
there are thrust faults and horizontal dislocations of internally undisturbed blocks.

Folds at the stoss end can be traced
using the thin lithified layer from the top
of gravels (Fig. 5, detail 1). The main
mass of gravel lies in cores of anticlines;
synclines are filled with sandy and sometimes gravelly diamicton. Sand layers
beneath the gravels appear to have acted as
shear planes.

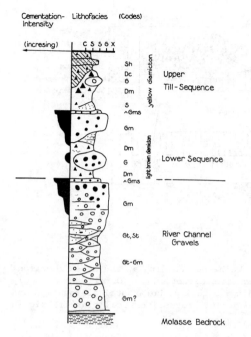

Fig. 4: Stratigraphic column of deposits at the Weihwang site. Intensity of carbonate cementation of gravels. Codes as in tab. 1.

Further towards the lee (= down ice direction), distinct blocks of internally more or less undeformed, crudely stratified and cemented gravels occur. The blocks are 5 - 10 m long and up to 1.8 m thick. The space between them is filled with sandy and gravelly material, sometimes diamicton.

A first block (Fig. 5, detail 2, block "1") is jointed near its stoss side and the fissure filled with cemented coarse sand. Beneath the block, there are more or less unconsolidated layers of sand and gravels, the latter with zones showing vertical clast fabric (pebbles reorientated). Horizontally elongated microfolds within the sands (diam. up to 10 cm, length >35 cm) lie just beneath the block forming its basal shear zone. They are also cemented with carbonate.

Beneath the next block (Fig. 5, detail 2, block "2"), unconsolidated sand and silt layers are up to 1.2 m thick. Near the proximal end, indistinctly striated and polished clay laminae are intercalated within silt, and may have acted as shear planes. The gravel block itself is thrust faulted in down ice direction, and along the thrust plane clasts are strongly reorientated.

The northernmost block (Figs. 5 and 6, block "3") as a whole dips slightly in a down ice direction. Its distal end is weakly folded, whereas the main part is undisturbed and still shows crude primary bedding structures and some distinct cross bedding near the bottom.

DISCUSSION: Summarizing all this, parts of the gravels reacted as competent blocks whereas other nearby elements of similar lithology were internally deformed (i.e. reacted as incompetent material). It is apparent that carbonate cementation frequently occurs in elements which behaved in a competent way, but is rarely found where internal deformation occured. One may conclude that cementation predates deformation. Were this the case, a phase of cementation would have to be introduced between deposition and dislocation. However, it is widely accepted that the time needed for cementation is quite considerable (deposits of the last glaciation are usually uncemented). Introduction of a cementation phase thus would mean not just a short episode but a severe alteration of the stratigraphic system. As there is no other evidence for it yet known, it seems unlikely that such a phase existed.

If carbonate cementation only happened during or after dislocation, it has to be discussed whether another kind of consolidation immediately postdating deposition needs to be postulated, in order to explain the block-like behaviour of the gravels during dislocation. In the first place, the subglacial behaviour of the material at a drumlin lee has to be considered. In this position, strata often are layered more or less conformable to the surface (which can be interpreted in different ways, e.g. primary bedding, "lee stratification", e.g. Dardis & McCabe 1987, or lee side shear structures, e.g. Menzies 1989). Conformable orientation of any gravelly intercalations would not be surprising in this environment, nor would a minor amount of horizontal dislocation (esp. as some of the till layers are identified as "lodgement" deposits). However it seems impossible that uncemented gravels become thrust faulted and jointed without acquiring any internal deformation features (the gravel blocks at Weihwang are undeformed).

This leads to the suggestion that (a) the blocks were frozen, i.e. cemented by an ice matrix during dislocation, and that (b) just those parts (blocks) which were frozen and dislocated are now consolidated by carbonate cement. "(b)" implies a mechanism of the carbonate cementation process which is somehow related to the melting of the ice

119

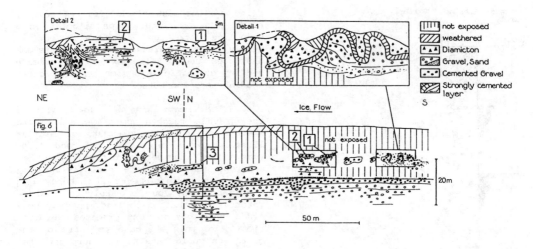

Fig. 5: Generalized longitudinal/cross section of Weihwang. Folding at the stoss, horizontal dislocation and stratification towards down ice direction. - Detail 1: Strongly cemented = lithified gravel stratum outlines folding. Detail 2: Horizontal dislocation and thrust folding of internally undisturbed cemented gravel blocks. Vertical clast fabric in uncemented gravels beneath right block.

matrix (here the different amounts of dissolved carbonates and other soluble components in meltwater versus nonglacial groundwater may play a part). - A similar situation where formerly frozen, dislocated gravel blocks are now cemented by carbonate whereas surrounding gravels of the same age are not, has been observed in a formerly fluvial environment (Verderber 1992).

A different mechanism of consolidation is considered for the lithified layers on the gravel top (Gms in Fig. 4). They mainly occur beneath matrix rich consolidated diamicton which can be identified as typical lodgement till. During their deposition, basal shearing restructured and consolidated this part of the gravelly, probably frozen, outwash.

3.2.2 Lithology of Marginal Till / Sand Sequences

Deposits at the lee margin are shown in Fig. 6. They consist mainly of layers of diamicton and sand. There are two main sequences, both starting with diamicton with intercalated sand and some proximal gravels, followed by layers of sand and fine material.

The basal unit again starts with light brown diamicton, similar to what occurs at the centre of the pit. It wedges out towards the margin of the landform.

At the level of dislocated gravel blocks

a transition zone follows, where layers of (still light brown) diamicton and of medium grained sand are interbedded. Pebbles in the proximal parts of two diamicton layers are orientated in a wide range of directions: some clasts are vertically orientated, and maxima in the ice flow direction (and oblique to it) are weak. One of the diamicton layers is elongated in boudin-structures, thinning out towards the margin (boudinage of diamicton within sand).

Again at the proximal side, at this level, gravelly material is incorporated into the sequence of sand and diamicton. There is a distinct south dipping lens of finer gravels, and a fold-like structure of coarser gravels. As similar material occurs in the adjoining megablock, debris may have been derived from there.

Further up, almost 5 m of monotonous, medium grained sand follow. The lower parts of it show distinct horizontal bedding (deformed with a slight southerly dip at the proximal and distal end). Next, a northward dipping yellow diamicton layer follows which, according to it's colour should be classified with the upper diamicton. The upper two meters of the sand appear indistinctly bedded, almost massive.

Beneath the diamicton layer, a number of couplet-like "clastic veins" were observed within the sand. They consist of almost pure silt of light brown and yellow colour. Some of them are conformable to bedding, whilst others are not. In the lower parts

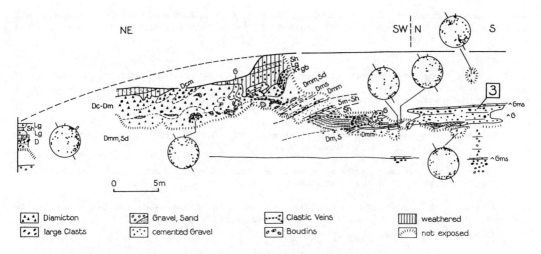

Fig. 6: Cross section at the right lee margin of the Weihwang drumlin (detail of fig. 5). Pole diagrams of clast fabric in diamicton. Codes as in tab. 1.

of the sands, a fold is outlined by a boudinaged silt vein. The maximal thickness of single veins is 3-4 cm at the proximal side, it diminishes towards the distal side, where about 2 mms were observed. - The veins are supposed to be intrusional features forced into the sand after deposition. Thinning out would indicate the direction of squeeze.

Above the sands, the yellow diamicton facies predominates. Lower parts are well-consolidated and matrix supported, occasionally with an indistinct pebble pavement at the bottom. The lower 1-2 m consist of almost pure diamicton, first massive, then indistinctly layered. Further up, the number of lenses and thin intercalations of coarse to fine sand and silt increases. Sand boudins within the diamicton are frequent. Pebble orientation beneath a circular structure (with gravel-dominated material in the centre) shows a very weak NNW maximum parallel to ice flow, and a more distinct ENE maximum towards the drumlin margin.

Towards the proximal end of the section, two coarse, clast-rich zones are incorporated into the diamicton. There is a vertical zone of about 2 m width where nests and clusters of matrix-poor gravels occur, along with a cluster-like fold, outlined with boulders up to 0.5 m and filled with coarse material. At the base of the latter a distinct but laterally limited pavement of large boulders commences. - The sequence is capped by decalcified layers of diamicton and sand.

Towards the margin, the diamicton grades up into increasingly clast rich and less consolidated facies. There are up to 4 m of this material showing only a few distinct structures (e.g. folds, boudins). The outward dip of strata, as indicated by indistinct layers of silt and coarse sand, conforms to the landform's surface.

An isolated section at the very northeast margin of the landform exhibits unconsolidated diamicton with horizontal and oblique sand layers. Large clasts dip NE-NNE; some are enveloped by silt and fine sand. Smaller pebbles clearly dip towards the margin of the landform. Some vertical to obliquely orientated fissures are filled with sand or coarser material.

DISCUSSION: Summarizing again, there are two till sequences at the lee margin of Weihwang drumlin. They both start with a layer of deformed (dragged, sheared) older deposits, followed by a diamicton appropriate to the definition of lodgement till sensu Dreimanis (1989). In the lower sequence, sands and gravels follow which are deformed or dislocated beneath the upper lodgement till. They sit more or less immediately on the lower diamicton. In the upper sequence, the diamicton facies changes gradually, indicating a continuing till sequence, with debris originating from sub-, en- and supraglacial positions. Layering of strata seems related to an already existing landform. There are more intercalations of gravels, and the number of shear structures decreases upwardly.

Depositional sequences and varying intensity of deformation make it unlikely that

Tab. 1: Abbreviations of lithology used in Figs. 4, 6, 7 (codes modified after Eyles et al. 1983, Miall 1985, Symbolschlüssel Geologie 1975/1991). Genetic interpretations are, where possible, in the text.

1. D = diamicton; Dm = matrix rich diamicton; Dc = matrix poor diamicton; Dmm, Dms = massive, stratified matrix rich diamicton; Dcm, Dcs = massive, stratified matrix poor diamicton; Lg = weathered, decalcified diamicton

2. G, -G, = gravel, lithified gravel; Gm = massive gravel; -Gms = .. sheared and lithified; Gt = stratified gravel in minor channels; fG, mG, gG (vb2) = fine, medium, coarse gravel (cemented); X, gb = large clasts, boulders

3. S, Sd = sand, deformed sand; Sm = massive sand; Sh = horizontally bedded sand; St = stratified (pebbly) sand in minor channels

4. F = fines; Fl = horizontally bedded (laminated) silt/fine sand; lam.U+S = vertically laminated silt and sand; u, fs = silt, fine sand (auxiliary constituent)

the formation of the Weihwang drumlin can be derived from only one of either processes. Stratigraphically, at least two glacier advance impulses were involved after deposition of the river channel gravels (Fig. 4).

During an early advance period, reshaping of the gravel surface took place: this is considered to mark the initial phase of drumlin formation, but deposition of a sequence starting with light brown diamicton also occured. The early advance is not represented by a complete till sequence. Some of the gravel intercalations which follow above the lodgement till may also have been deposited in a subglacial environment, probably under stagnant conditions. The other part is suggested to be (proglacial) outwash.

During a later advance period, layers of frozen gravels and of mobile material occured between the active ice and the completely frozen, immobile underground (sensu Menzies 1989). In this environment, parts of the frozen gravels were dislocated, and they became redeposited as distinct gravel blocks, whereas adjoining deposits were folded, indicating more mobile conditions. The latter are also indicated by the intrusion of clastic silt veins. All this happened, however, only within distinct layers of the sequence, whilst the general stratification was still preserved.

4   THE HIRSCHBRUNNEN DRUMLINS

The Bodanrück drumlin field, especially the Hirschbrunnen drumlins near Radolfzell / Markelfingen (site 3, Fig. 1), has become one of the key localities in the study of drumlin formation processes in the Rhine glacier area (Habbe 1988, 1989, Ellwanger 1990). As at Weihwang, the lithological sequence consists of undisturbed gravels beneath deformed till of various thicknesses. Additionally clastic dykes and sills have been squeezed into the gravels. Generally these dykes are younger than the adjoining gravels but they predate (or are contemporanious with) deposition of the upper till. They can be used to identify dislocations of parts of the gravels during till deposition, and to subdivide deposition and deformation of the till.

There are three elements in the stratigraphic sequence at the Hirschbrunnen drumlin (Fig. 7): a "lower unit" of consolidated fines with intercalated matrix-rich diamicton layers (lacustrine and glaciolacustrine deposits), a "middle unit" which is gravel dominated with some diamicton intercalations (kame terrace deposits, in which dykes and sills have been observed), and an "upper unit" consisting of diamicton and reworked material (drumlin till of widely varying thickness, related to the surface of the underlying gravels). Three places where dykes and sills intrude into the gravels will be described in detail.

Without exception, the dykes and sills consist of laminated silt and fine sand. However, lens-like zones of fine-grained diamicton, with clast sizes up to 2 mms were are also found in some dykes. On average, this facies is 15 - 30 cm thick, apart from distal parts of the sills where they wedge out (5 - 10 cm). Gravels and till (diamicton) are also incorporated into the "Great Dyke", and as a result it becomes up

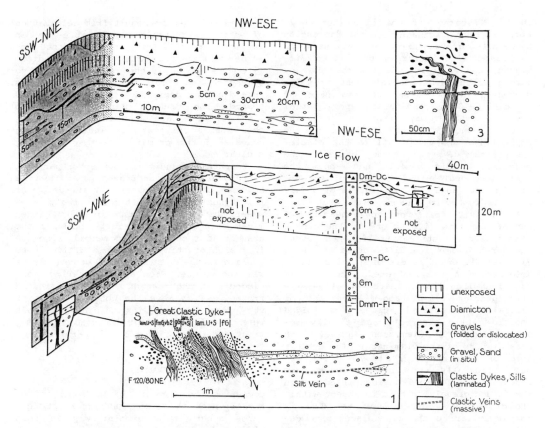

Fig. 7: Longitudonal and cross section and stratigraphic column of Hirschbrunnen drumlins near Radolfzell / Markelfingen. Details 1-3 show different types of dislocations of gravels as recognizable from vertical clastic dykes and horizontal clastic sills.

to 120 cm thick (with the vertically laminated fines still 15 - 30 cm thick). - Another non depositional feature is small clastic veins of massive silt with no or only indistinct laminae. These are up to 4 cm thick and occur not only in gravels as couplet-like intercalations, but also within dykes.

The varying exposures of the "Great Dyke" are situated in the SW corner of the site. However the dyke itself is known to run for 4-5 km parallel to the shore of a deeply eroded branch basin of Lake Constance (Ellwanger 1990). It strikes parallel to ice flow and to drumlin long axes. Fig. 7 (detail 1) shows the 1991 exposure. Here the dyke appears subdivided with intermittant gravel blocks, and a vertical component of dislocation is indicated by a sandy intercalation in adjoining gravels, with an uplift of about 20 cm at the side of the

basin. Veins of non-laminated clay and silt can be observed within the dyke itself and along the sand intercalation, the latter is 3-4 cm thick proximal to the dyke but decreasing to <1 cm in further distal zones. Veins within the dyke are about 1 cm thick. They are clay-rich and indistinctly laminated.

The fine material of the "Great Dyke" is similar to the glaciolacustrine deposits from beneath the gravels. It would seem that the dyke material intruded upward. Coarser parts of the Great Dyke may either have more or less remained in place or - if the gravel facies is unlike adjoining gravels - have been deposited by gravity. Some lenses of coarse sand may also have been horizontally transported. The veins represent a late phase of intrusion as they cut both gravels and dykes.

Horizontal and oblique sills are exposed

in the NW corner of the site (detail 2, Fig. 7). They originate from the depression between two gravel surface elevations which is filled with leewardly stratified diamicton, and they intrude towards the lee margin of the drumlin. Thickness of sills decreases in down ice direction (max. 30 cm, min. 5-10 cm), indicating that material was squeezed that way (down ice). Material appears similar to the diamicton of the upper unit, and it is unlike the glaciolacustrine deposits from beneath the gravels. Gravels beneath and above the sills seem undisturbed. At one of the oblique parts, an uplift of gravels of 5 - 10 cm is indicated by a sandy intercalation.

A minor vertical "cross" dyke near the SE end of the site (i.e. close to the centre of the drumlin) strikes at right angles to the direction of ice flow (detail 3, Fig. 7). Here different levels of glaciotectonic deformation are illustrated. The gravels are in situ (lower part of fluvially structured gravels), dragged (upper part of fluvially structured gravels), folded (pebble fabric altered), and finally reworked into the diamicton (gravel rich basal layers of the diamicton). The dyke is more or less undisturbed where it cuts still fluvially structured gravels, and traces of it can still be identified within the folded gravels, but it is absent within the diamicton. - Material of the dyke is again similar to the glaciolacustrine deposits from beneath the gravels.

DISCUSSION: In longitudinal section (Fig. 7, ESE-NW), both cross dyke and sills are situated in places where the gravel surface is elevated and only just veiled by a cover of more or less primarily deposited till. Here only a thin mobile layer between ice and rigid basement can have existed, and the different dislocations illustrate that subglacial erosion of the gravel surface predominated. Between the two elevations, in the "sheltered" depression of gravel surface, leewardly dipping strata of diamicton (maybe the only upper part of them) may have acted as a thicker mobile layer. Here primary deposition can be considered, as "lee stratification" (e.g. Dardis & McCabe 1987) or "cavity infill" (e.g. Shaw et al. 1989), but conditions during deposition and/or dislocation must also suffice to sqeeze mobile fine material into the leewardly adjoining sills.

In the cross section (Fig. 7, SSW-NNW), layers of till dipping towards the drumlin margin are recognizable immediately above the gravel surface too. However, the structures as a whole appear as inclined folds dipping towards the landform margin.

This is believed to result from deformation of stratified till (conformable to an earlier surface), subglacially turned upwards towards the crest of the landform (i.e. as part of the general down ice transport, material is removed from the interdrumlin depression and redeposited towards the drumlin margin, cf. Aario 1987). Again a late phase of deformation in this mobile substrate is indicated, with genesis of the primary layering remaining unclear.

The question again rises whether parts of the gravels had to be frozen (i.e. cemented by ice) during fracture and intrusion of the dykes and sills. It seems likely, however, that parts of the intruding fines might escape into pore-rich zones of the gravels if those were uncemented. Also the intrusion of thin veins seems difficult to explain if the matrix between gravel clasts was incomplete. From all this it is concluded that the mass of gravels was indeed frozen. - Only when the uppermost part of the cross dyke and its adjoining gravels were folded (i.e. after intrusion), permafrost did retreat from this upper zone.

5. SUMMARY, CONCLUSIONS

Comparing the drumlins of Weihwang and of Markelfingen it seems that important features occur in a similar way in both places:

(I) the strongly eroded gravel surfaces beneath the landforms,

(II) the horizontally dislocated but internally undisturbed "competent" gravel blocks,

(III) the lee-side stratified sheets of till more or less conformable with the gravel and/or landform surfaces; and finally

(IV) the intrusional features, which, combined with glaciotectonic structures, indicate a mobile state of part of the substrate during some of the deformation phases.

The sequence of processes seems to be more or less similar in both places.

One may therefore conclude that even if the morphology of Riss-aged drumlins is less distinct than those of Würm age, they still both originate from similar conditions. The flat shape of Riss drumlins may therefore indeed be caused by their geographic position outside the main glacial basin where, taken at a large scale, the bedrock surface dips in direction of ice flow.

References

Aario, R. 1987: Drumlins of Kuusamoo and Rogen ridges of Ranua, northeast Finland. - Drumlin Symposium, ed. Menzies J. & J. Rose 87-101, Rotterdam (Balkema).

Dardis, G.F. & McCabe, A.M. 1987: Subglacial sheetwash and debris flow deposits in late-Pleistocene drumlins, Northern Ireland. - Drumlin Symposium, ed. Menzies J. & J. Rose, 87-101, Rotterdam (Balkema).

Dreimanis, A. 1989: Tills: Their genetic terminology and classification. - Genetic Classification of Glacigenic Deposits, ed. Goldthwait, R.P. & C.L. Matsch, 17-84, Rotterdam (Balkema).

Ellwanger, D. 1990: Würmzeitliche Drumlin-Formung bei Markelfingen (westlicher Bodensee, Baden-Württemberg). - Jber. Mitt. oberrhein. geol. Ver., N.F., 72, 411-434; Stuttgart.

Eyles, N., Eyles, C.H. & Miall, A.D. 1983: Lithofacies types and vertical profile models; an alternative approach to the description and environmental interpretation of glacial diamict and diamictite sequences. - Sedimentology, 30, 393-410, Amsterdam.

Grube, F. 1980: Zur Morphogenese und Sedimentation im quartären Vereisungsgebiet Nordwestdeutschlands. - Verh. naturwiss. Ver. Hamburg, (NF)23, 69-79, Hamburg.

Habbe, K.A. 1988: Zur Genese der Drumlins im süddeutschen Alpenvorland - Bildungsräume, Bildungszeiten, Bildungsbedingungen. - Z. Geomorph. N.F., Suppl.-Bd. 70, 33-50, Berlin - Stuttgart.

Habbe, K.A. 1989: On the origin of the drumlins of the South German Alpine foreland. - Sediment. Geol., 62, 357-370, Amsterdam.

Miall, A.D. 1985: Glaciofluvial Transport and Deposition. - In: Eyles, N. (ed.), Glacial Geology, 168-183, Oxford.

Menzies, J. 1987: Towards a general hypothesis on the formation of drumlins. - In: Menzies, J. & Rose, J.(eds): Drumlin Symposium, 9-24, Rotterdam/Boston (Balkema).

Menzies, J. 1989: Subglacial hydraulic conditions and their possible impact upon subglacial bed formation. - Sediment. Geol., 62, 125-150, Amsterdam (Elsevier).

Penck, A., & Brückner, E. 1901/09: Die Alpen im Eiszeitalter, Bd. 1-3, Leipzig (Tauchnitz).

Schreiner, A. 1973: Erläuterungen zu Blatt 8219 Singen. - Geol. Kt. Baden-Württ. 1:25000, Stuttgart.

Schreiner, A. 1974: Erläuterungen zur Geologischen Karte des Landkreises Konstanz mit Umgebung 1:50000, Stuttgart (Landesvermessungsamt Baden-Württ.).

Schreiner, A. 1989: Zur Stratigraphie der Rißeiszeit im östlichen Rheingletscher-Gebiet (Baden-Württemberg). - Jh. geol. Landesamt Baden-Württ., 31, Freiburg.

Schreiner, A. & Haag, T. 1982: Zur Gliederung der Rißeiszeit im östlichen Rheingletschergebiet (Baden-Württemberg). - Eiszeitalter u. Gegenwart, 32, 137-161, Hannover.

Shaw, J., Kvill, D. & Rains, B. 1989: Drumlins and catastrophic subglacial floods. - Sediment. Geol., 62, 177-202, Amsterdam.

Symbolschlüssel Geologie 1975/1991. - Hrsg. von den Geologischen Landesämtern in der Bundesrepublik Deutschland, Hannover. (2. Aufl. 1975, Fassung 1991)

Verderber, R. 1992: Quartärgeologische Untersuchungen am Hochrhein. - Abh. Geol. Landesamt Baden-Württ., 14, Freiburg.

Werner, J. 1975: Erläuterungen zu Blatt 8020 Meßkirch. - Geol. Kt. Baden-Württ. 1:25 000, Stuttgart.

Formation and Deformation of Glacial Deposits, Warren & Croot (eds) © 1994 Balkema, Rotterdam, ISBN 90 5410 096 6

Late-glacial resedimentation of drumlin till facies in Ireland

George F. Dardis
Sedimentology and Palaeobiology Laboratory, Anglia Polytechnic, Cambridge, UK

Patricia M. Hanvey
Department of Geography and Environmental Studies, University of the Witwatersrand, Johannesburg, South Africa

Pete Coxon
Department of Geography, Trinity College, Dublin, Ireland

ABSTRACT: Several genetic varieties of resedimented diamictons from three major depositional environments (subaerial, sublacustrine, periglacial-marine) occur on the surfaces and flanks of drumlins of late Pleistocene age in Ireland. Six sites date the resedimentation events to the late-glacial period (15 - 10 ka B.P.) and later, suggesting that resedimentation occurred in several phases during and following deglaciation. Differentiation of these facies types from undisturbed diamicton facies formed either prior to or during drumlinization is important as much of the evidence concerning drumlin-streamlining processes are associated with sedimentary facies in the drumlin surface carapace.

1. INTRODUCTION

The widespread formation of drumlins in areas of former glacial cover in the northern hemisphere is one of the major enigmas surrounding ice sheet behaviour during the last cold stage. Increasingly, these landforms are seen to contain vital sedimentary evidence of changes in the nature and pattern of subglacial sedimentation during the course of a glacial cycle (Dardis 1985). This evidence can, for example, be used to examine changes in subglacial hydrology over time (Shaw 1983, Dardis et al. 1984), which may be an important mechanism controlling surging behaviour of ice sheets (Dardis et al. 1984, Shaw 1989) and promoting rapid glacial terminations (McCabe & Dardis 1989a, Shaw 1989).

Much of this sedimentary evidence has been correlated morpho-stratigraphically. Sedimentary facies contained within the drumlin construct have, for example, been related to pre-streamlining (or pre-drumlinization) events, while those associated with the drumlin carapace, or requiring a proto-drumlin construct to allow their formation, have been related to streamlining (i.e. drumlinization) and subsequent events (Dardis 1985, Dardis et al. 1984). When using this approach, it is important to have clear and unambiguous information about the ages and palaeoenvironmental significance of drumlin carapace sedimentary facies (Dardis and McCabe 1983, Dardis 1985).

Here we present evidence from Ireland of resedimentation of the surficial layers of drumlins which occurred during the late-glacial period (15-10 ka BP) and later. These facies are significant as they often lack clearly discernible evidence of remobilisation and pin-point potential difficulties in differentiating between true glacigenic diamictons and those which have undergone resedimentation in drumlin stratigraphic sequences.

2. LOCALITIES

Several thousand drumlins have been examined in detail in south-central Ulster (Dardis 1982) and western Ireland (Hanvey 1988). Of these six show sedimentary facies which can be clearly associated with post-drumlinization sedimentation events (Fig. 1).

Fig. 1. Location map (1: Ardgivna 2: Curran 3: Seahill 4: Port Inver 5: Old Head 6: Moneymore).

2.1 *Ardgivna*

A complete drumlin and inter-drumlin hollow sedimentary sequence is exposed in a roadstone quarry at Ardgivna (Site 1, Fig. 1). The drumlin is composed of a clay-rich diamicton unit, 5-30 m in thickness, overlying a rock basement consisting of glacitectonized heavily shattered Carboniferous mudstones (Dardis 1982). The inter-till hollow sequence consists mainly of alternating beds of sands, silts and clays with intraformational diamicton units (Fig. 2).

The sand/silt/clay sequence consists of massive sands, type-A and type-B ripple-drift cross-laminations, type-S laminations and mud drape laminations (Jopling & Walker 1968, Southard et al. 1972) (Fig. 3). Facies relationships within these deposits, while showing transitions from bedload to suspended load sedimentation, similar to vertical sequences found in glaciolacustrine bottomset sequences (cf. Ashley 1975, Gustavson et al. 1975), also

Fig. 2. View of the laminated sequence, Ardgivna inter-drumlin hollow.

show unusually high proportions (70%) of current-dominated sand facies. This probably reflects the unusual depositional niche (i.e. inter-drumlin hollow), where sublacustrine water was channelled by the highly undulating drumlin topography, thereby imparting fluviatile affinities to bottomset sequences by convergence of bottom-flowing water. The upper limit of the bottomset sequences at Ardgivna occurs at 77 m I.O.D. (Irish Ordnance Datum), which is ~30 m below local late Quaternary glacial lake levels (Dardis 1986).

The diamicton unit within the glaciolacustrine sequence is 0.2-2.0 m thick and 27 m long, and displays considerable lateral variations in thickness (Fig. 3). The major portion of the lacustrine sequence is draped over the diamicton unit, with only a thin (30-50 cm) basal unit lying beneath it, consisting of parallel-laminated sands with interbedded mud drape laminae (Fig. 4). Within the diamicton unit there is an upper component which shows inverse-to-normal grading and lateral changes in grading, and a lower component characterized by well-defined fissility at the base. Small cobble-sized diamicton intraclasts (Tc; Fig. 5) occur in the lacustrine sediments underlying the diamicton

Fig. 3. Lensate diamicton unit, Ardgivna inter-drumlin hollow.

unit, suggesting some degree of fragmentation of the unit during emplacement.

Macro-fabrics in the diamicton units show V_1 values of 270°/25°, as opposed to 330°/15° in basal tills in neighbouring drumlins (Dardis

Fig. 4. Detailed view of the upper and lower boundaries of the intraformational diamicton unit, Ardgivna inter-drumlin hollow. (Note the trowel for scale at the base of the unit).

1982), showing that flow occurred in a direction transverse to ice flow under the influence of the local topography (i.e. drumlins). This and the textural characteristics of the unit suggests that the sediment mix was derived from the adjacent drumlin flank.

The morphological and textural characteristics of the diamicton unit, showing grading, localized basal shearing and relatively good preservation of the original sediment mix, is typical of high density, undersaturated debris flows (Johnson, 1970). However, this contradicts the intraformational nature of the unit, which suggests formation in a subaquatic environment, where there is high potential for rapid fall-off in competence (Cohen 1983, Hampton 1975, Eyles 1987, Eyles et al. 1987). This dichotomy is explained by (1) the very high mud content (~90 %) of the parent material (glacigenic diamictons composed almost entirely of mudstones) and the relatively low-energy depositional environment, which may have inhibited dilution of the flow unit (cf. Hampton 1972), and (2) the basal sheared layer, which suggests flow emplacement by a conveyor mechanism which would have resulted in relatively little disturbance of underlying sediments and passive rafting of much of the material in transit (cf. Rodine & Johnson 1976, Lowe 1982).

Fig. 5. View of the lower boundary of the intraformational unit, Ardgivna inter-drumlin hollow. (Tc: diamicton intraclast in the laminated sequence).

2.2 *Curran*

A small section in a disused brick-pit at Curran (Site 2, Fig. 1) shows a transverse exposure, 40 m long and 2-5 m high, on the flank of a drumlin, comprising a laminated silt and clay sequence underlain and overlain by diamicton lithofacies.

The lower diamicton is massive and clay-rich; clasts within the diamicton are generally well striated and the unit appears to be *in situ*.

The laminated sequence consists mainly of 2 m of parallel-laminated silts and clays, showing high proportions of evenly bedded mud/silt couplets, similar to proximal-to-distal glaciolacustrine rhythmites (cf. Ashley 1975).

The laminated sequence is overlain by a stratified diamicton facies, consisting of massive diamicton units, 20-30 cm thick, interbedded with thin (~5-10 mm) mud laminae. The latter are not as well sorted as in the laminated sequence. While facies of this type can sometimes form subglacially (cf. Dardis & McCabe 1983, 1987), it is most likely that they reflect subaerial deposition from highly saturated sediments, by flow in rills and sheets over the diamicton unit during successive resedimentation phases (cf. Lawson 1982).

2.3 *Seahill*

Realignment of the main Donegal to Killybegs road resulted in 360m of exposure in an east-west road cutting towards the northern (stoss) end of a drumlin (Site 3; Fig. 1).

These exposures reveal lateral facies transition from massive, compact limestone/shale-rich diamictons (logs B,C,G,H; Fig. 6) to diamictic muds. The diamictic muds form the basal component in a number of logs (logs D-F; Fig. 6), but are also interbedded with massive diamictons and minor sand and silt units (logs D,E; Fig. 6). The diamictic muds are similar to sedimentary facies found as basal components of drumlins elsewhere in western Ireland (McCabe & Dardis 1989b) and in morainal complexes in eastern Ireland (McCabe et al. 1984, 1987) which are interpreted as glaciomarine.

An extensive organic layer with a maximum thickness of 23 cm occurs within the road cutting (logs A, I; Fig. 6) and is continuously exposed laterally for ~50 m in the northern flank of the road cutting and for ~20 m on the southern flank (see Fig. 6). At one point (log I; Fig. 5) the organic horizon is overlain by 5 m of massive diamicton. The organic horizon consists of a dark brown predominantly fine sand and silt, with subangular quartz grains, with 4 per cent organic carbon material. The organic material contains very fine plant detritus and some charcoal fragments. Tiny (<2mm) gastropod shells were extracted from the organic horizon but these have not yet been identified or dated.

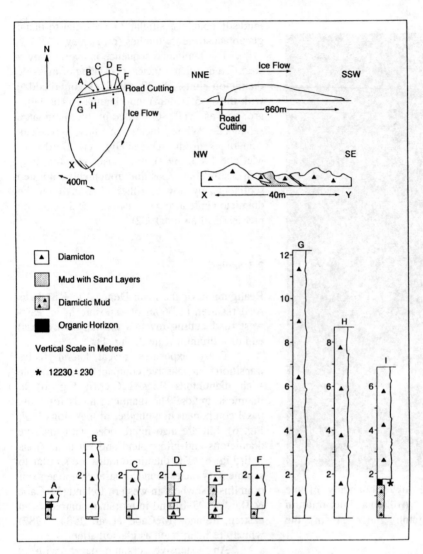

Fig. 6. Morphology and sedimentary logs, Seahill drumlin.

The pollen concentration in the organic layer was low (< 3500 grains cm^{-3}) and prevented detailed analysis. However, the assemblage (Table 1) is typical of the colder phases of the late-glacial period. It is unlikely that the organic materials accumulated during warmer parts of the Late-glacial as marker taxa for these phases are absent (e.g. *Juniperus*). This interpretation is supported by a radiocarbon date of 12,230 ± 230 years B.P. (Beta-24804), which places the organic layer within the Woodgrange Interstadial (cf. Watts 1977).

2.4 Port Inver

A drumlin at Port Inver (Site 4; Fig. 1) has been exposed by coastal erosion and shows an extensive longitudinal section 500 m in length, which is close to the central crest line of the drumlin (Fig. 7).

This drumlin comprises a lower compact, limestone/shale-rich diamicton and rests on a striated shale bedrock, which grades laterally into diamicton regolith, consisting of locally derived angular clasts. The lower till facies is

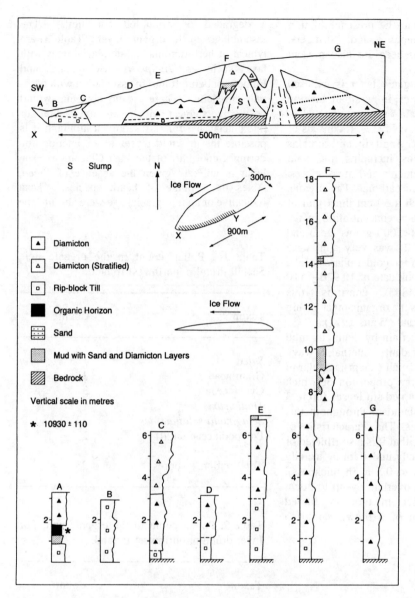

Fig. 7. Morphology and sedimentary logs, Port Inver drumlin.

overlain by a crudely stratified, sandstone-rich diamicton. The basal diamicton thins leewards and is relatively homogeneous throughout except for the occurrence of a single clast boulder lag (Fig. 7). It is separated from the overlying sand-rich diamicton by an irregular, silty-sand unit, which has a maximum thickness of 2.5 m but this varies greatly because of a very irregular basal boundary (log f, Fig. 7). The latter is characterized by thin, downward finger-like projections (up to 1m in length) which penetrate into underlying sediments. Small irregular blocks of the lower diamicton are incorporated into the mud/sand unit. The upper boundary of the Fm/Sm lithofacies, in contrast to the lower junction, is distinct but regular, and dips gently (~7°) to the southwest (leewards). The upper diamicton has a sandy texture, and it displays

133

crude stratification due to the presence of thin (<0.5 cm) discontinuous sand stringers. Occasional thicker sand beds are also present (log f, Fig. 7).

A 4 cm thick organic layer is exposed over a distance of 4 m in a small section perpendicular to the main exposure at the lee-end of the drumlin (log A; Fig. 7). It consists of dark brown/black fine sand and silt, and contains abundant organic detritus, including fine plant materials, charcoal fragments and moss spores, with 30 % total organic carbon. The organic layer is interbedded with a 60 cm thick unit of massive silts. The pollen concentration in the organic layer was low (<5000 grains cm^{-3}) and the assemblage (Table 2) was very similar to that at Seahill, typical of the colder phases of the late-glacial. A radiocarbon date of 10,930 $\pm$ 110 years B.P. (Beta-24803) confirms this interpretation and places the organic layer within the Nahanagan Interstadial (Watts 1977).

The silts are underlain by diamictic mud which displays abundant internal, wavy lamination. It contains small (<5cm), angular to subangular clasts (a large proportion of which are rotted). Small, disrupted silt lenses (up to 3 cm in thickness) are abundant throughout this diamictic mud lithofacies. The characteristics of this lithofacies suggests that it is a solifluctued or flow deposit. It is directly underlain by heavily weathered bedrock (up to 0.5 m in thickness). The organic horizon is overlain by up to 30cm of massive silts, which in turn are buried beneath at least 1.6 m of a brown sand-rich diamicton (Fig. 7).

2.5 Old Head

A coastal section at Old Head (site 5; Fig. 1) show an extensive exposure of sediments along a transition zone between a drumlin and a stoss-side megacavity infill (Hanvey & Dardis, unpublished). An exposure in the western part of the exposure shows two organic layers underlying ~2 m of massive till. A similar sequence at Old Head which was thought to be interstadial or interglacial in age was described by Synge (1970).

Both organic layers showed low pollen concentrations, with many of the pollen grains in a degraded or crumpled condition. The assemblages in the organic layers (Table 3) are typical of herb-dominated late-glacial ones with taxa such as *Juniperus*, Selaginella and Empetrum being found in association with herb-rich grassland. There are no taxa that would positively suggest either an interstadial or interglacial age for this sediment although it is possible that it could represent an amelioration during a cool part of the Last Glaciation. The latter is unlikely given the absence of *Picea*, Abies or a range of heath species. Taxa indicative of an interglacial age were absent. The

Table 1. Pollen count in the organic unit, Seahill drumlin, northwest Ireland.

Taxon	Count*
Salix	1
Gramineae	3
Cyperaceae	3
Thalictrum	1
Selaginella selaginoides	13
Polypodiaceae undiff.	2
Total pollen + spores	23

Table 2. Pollen count in the organic unit, Port Inver drumlin, northwest Ireland.

Taxon	Count*
Pinus	1
Gramineae	3
Cyperaceae	3
Ericales undiff.	1
Rosaceae	1
Rumex	1
Selaginella selaginoides	6
total pollen + spores	16

Table 3. Tree, shrub and herb palynomorphs* in the organic units at Old Head, western Ireland.

	Unit 1	Unit 2
Betula	3	14
Pinus	1.5	7
Juniperus	5	
herbs		
Gramineae	31.5	36
Cyperaceae	51	17
Empetrum	0.5	
Compositae (tub)	4	
Caryophyllaceae	0.5	7
Cruciferae	1.5	7
Ranunculaceae	0.5	
Thalictrum	2	
Umbelliferae	5	5
Sphagnum		1
Polypodiaceae		2
Lycopodium	37	16
Selaginella	12	12
Pollen sum (P)	133	42
organic detritus	+++	+++

* Tree, shrub and herb palynomorphs expressed as a percentage of the pollen sum (P). Lower plants as a percentage of P + lower plants.

Fig. 8. Clast-rich layer marking the Late-glacial/Post-glacial transition, Moneymore inter-drumlin hollow.

assemblages are similar to those found between 11,000 and 10,000 years B.P., suggesting that this is probably the same sequence described by Mitchell (1977) at Old Head, which yielded an radiocarbon date of 10,000 years B.P.

2.6 Moneymore

An extensive drumlin and inter-drumlin hollow sedimentary sequence is exposed in a quarry at Moneymore (Site 6; Fig. 1). It shows complex drumlin flank resedimentation, with a deep (~3-4 m) infills, in the inter-drumlin hollow, of poorly-sorted sands of fluvioglacial origin interbedded with diamicton units. The sands are overlain by a late-glacial and post-glacial mud-peat

sequence, showing a distinct clast-rich zone (Fig. 8) which marks the late-glacial/post-glacial transition (i.e. the Juniper rise) in pollen sequences at the site (Hirons 1980).

3. DISCUSSION AND CONCLUSIONS

(1) The sites demonstrate resedimentation during several phases following drumlin streamlining. The Ardgivna site clearly demonstrates extensive resedimentation during the Annahavil lacustrine phase of deglaciation which may have been coeval with the Armoy Stage at 14,000-13,500 years B.P. (Dardis 1986). Some of the early resedimentation at Moneymore, represented by diamictons interbedded with sands, may have occurred at about this time. The Moneymore site also demonstrates minor resedimentation during the late-glacial/post-glacial transition, associated with rapid climatic amelioration. The other sites

all demonstrate clear evidence of resedimentation, but the timing of resedimentation events is uncertain. It is likely, however, that they occurred sometime within the late-glacial period, when limited vegetation cover and rapid changes in ground conditions due to climate oscillations would have promoted mass movements.

(2) The extent of resedimentation, particularly at Seahill, raises questions about the degree of post-deposition modification that late Pleistocene drumlins have undergone. Drumlin flank resedimentation on the scale of the Seahill drumlin has been described in Scotland (Dickson et al. 1976), suggesting that it is probably relatively common but largely unrecognised.

(3) If resedimentation is widespread, it poses three major problems which need to be addressed. Firstly, do we have a clear idea of the morphology of late Pleistocene drumlins at their time of formation? Resedimentation data suggests that their form was probably more pronounced, with relatively steep lateral gradients. If this is the case, it has considerable implications in terms of models of drumlin genesis. Secondly, do we have clear information on primary depositional sequences within drumlins and how easy is it to distinguish between these and major drumlin flank resedimentation profiles (cf. Stea & Mott 1989)? This may not be possible if large parts of the drumlin (~30-50 %) have undergone resedimentation and suggests the need for greater caution when interpreting drumlin and inter-drumlin sedimentary profiles (cf. Boyce & Eyles, 1991). Thirdly, can we clearly distinguish between glacial diamictons and resedimented glacial diamictons in these profiles. At the present time, it seems that this is generally only possible where there is clear supplementary evidence of resedimentation.

4. ACKNOWLEDGEMENTS

We thank John Shaw, University of Alberta for helpful comments on a draft of the paper and Dr. Ken Hirons, University of Birmingham, who provided useful pollen data and field assistance (to G.F.D.) at the Moneymore inter-drumlin site.

5. REFERENCES

Ashley, G.M. 1975. Rhythmic sedimentation in glacial lake Hitchcock, Massachusetts-Connecticut. In Jopling, A.V. & B.C. McDonald (eds.) *Glaciofluvial and glaciolacustrine sedimentation*: 304-320. Tulsa, Society of Economic Palaeontologists and Mineralogists, Special Publication 23.

Boyce, J.I. & N. Eyles 1991. Drumlins carved by deforming till streams below the Laurentide ice sheet. *Geology*, 19: 787-790.

Cohen, J.M. 1983. Subaquatic mass flows in a high energy ice marginal deltaic environments and problems with the identification of flow tills. In Evenson, E.B., Schluchter, Ch. & J. Rabassa (eds.) *Tills and related deposits*: 255-267. Rotterdam, Balkema.

Dardis, G.F. 1982. *Sedimentological aspects of the Quaternary geology of south-central Ulster, Northern Ireland*. Unpublished Ph.D. thesis, Ulster Polytechnic. 433pp.

Dardis, G.F. 1985. Till facies associations in drumlins and some implications for their mode of formation. *Geografiska Annaler*, 67A: 13-22.

Dardis, G.F. 1986. Late Pleistocene glacial lakes in south-central Ulster, Northern Ireland. *Irish Journal of Earth Sciences*, 7: 133-144.

Dardis, G.F. & A.M. McCabe 1983. Facies of subglacial channel sedimentation in late Pleistocene drumlins, Northern Ireland. *Boreas*, 12: 263-278.

Dardis, G.F. & A.M. McCabe 1987. Subglacial sheetwash and debris flow deposits in late Pleistocene drumlins, Northern Ireland. In Menzies, J. & J. Rose (eds.) *Drumlin symposium*: 225-240. Rotterdam, Balkema.

Dardis, G.F., McCabe, A.M. & W.I. Mitchell 1984. Characteristics and origins of lee-side stratification sequences in late Pleistocene drumlins, Northern Ireland. *Earth Surface Processes and Landforms*, 9: 409-422.

Dickson, J.A., Jardine, W. & R.J. Price 1976. Three late-glacial sites in west-central Scotland. *Nature*, 262: 43-44.

Eyles, N. 1987. Late Pleistocene debris flows in large glacial lakes in British Columbia and

Alaska. *Sedimentary Geology*, 53: 33-71.

Eyles, N., Clark, B.M. & J.J. Clague. 1987. Coarse-grained sediment gravity flow facies in a large supraglacial lake. *Sedimentology*, 34: 93-116.

Gustavson, T.C., Ashley, G.M. & J.C. Boothroyd 1975. Depositional sequences in glaciolacustrine deltas. In Jopling, A.V. & B.C. McDonald (eds.) *Glaciofluvial and glaciolacustrine sedimentation*: 264-280. Tulsa, Society of Economic Palaeontologists and Mineralogists, Special Publication 23.

Hampton, M.A. 1972. The role of subaqueous debris flows in generating turbidity currents. *Journal of Sedimentary Petrology*, 42: 775-793.

Hampton, M.A. 1975. Competence of fine-grained debris flows. *Journal of Sedimentary Petrology*, 45: 834-844.

Hanvey, P.M. 1988. *The sedimentology and genesis of late-Pleistocene drumlins in Counties Mayo and Donegal, western Ireland.* Unpublished D.Phil. thesis, University of Ulster.

Hirons, K.R. 1980. Moneymore. In Edwards, K.J. (ed.) *IQUA field guide to Co. Tyrone*: Belfast, Queens University.

Johnson, A.M. 1970. *Physical processes in geology*. San Francisco, Freeman, Cooper and Co. 577 pp.

Jopling, A.V. & R.G. Walker 1968. Morphology and origin of ripple-drift cross-stratification, with examples from the Pleistocene of Massachusetts. *Journal of Sedimentary Petrology*, 38: 971-984.

Lawson, D.E. 1982. Mobilization, movement and deposition of active subaerial sediment flows, Matanuska Glacier, Alaska. *Journal of Geology*, 90: 279-300.

Lowe, D.R. 1982. Sediment gravity flows: II. Depositional models with special reference to the deposits of high-density turbidity currents. *Journal of Sedimentary Petrology*, 52: 279-297.

McCabe, A.M. & G.F. Dardis 1989a. A geological view of drumlins in Ireland. *Quaternary Science Reviews*, 8: 169-177.

McCabe, A.M. & G.F. Dardis 1989b. Sedimentology and depositional setting of late Pleistocene drumlins, Galway Bay, western Ireland. *Journal of Sedimentary Petrology*, 59, 944-959.

McCabe, A.M., Dardis, G.F. & P.M. Hanvey 1984. Sedimentology of a late-Pleistocene submarine-moraine complex, County Down, Northern Ireland. *Journal of Sedimentary Petrology*, 54, 716-730.

McCabe, A.M., Dardis, G.F. & P.M. Hanvey 1987. Sedimentation at the margins of a late Pleistocene ice-lobe terminating in shallow marine environments, Dundalk Bay, eastern Ireland. *Sedimentology*, 34: 473-493.

Mitchell, G.F. 1977. Periglacial Ireland. *Philosophical Transactions of the Royal Society of London*, B, 280: 199-209.

Rodine, J.D. & A.M. Johnson 1976. The ability of debris, heavily weighted with coarse clastic materials, to flow on gentle slopes. *Sedimentology*, 23: 213-234.

Shaw, J. 1983. Drumlin formation related to inverted meltwater erosion marks. *Journal of Glaciology*, 29: 461-479.

Shaw. J. 1989. Drumlins, subglacial meltwater floods and ocean responses. *Geology*, 17: 853-856.

Southard, J.B., Ashley, G.M. & J.C. Boothroyd 1972. Flume simulation of ripple-drift sequences. *Geological Society of America, Abstracts and Programs*, 4: 672.

Stea, R.R. & R.J. Mott 1989. Deglaciation environments and evidence for glaciers of Younger Dryas age in Nova Scotia, Canada. *Boreas*, 18: 169-177.

Synge, F.M. 1970. The Irish Quaternary: current views 1969. In Stephens, N. & R.E. Glasscock (eds.) *Irish geographical studies*: 34-48. Belfast, Queens University.

Watts, W.A. 1977. The Late Devensian vegetation of Ireland. *Philosophical Transactions of the Royal Society of London*, 280B: 273-293.

Interpreting the glacial record

Formation and Deformation of Glacial Deposits, Warren & Croot (eds) © 1994 Balkema, Rotterdam, ISBN 90 5410 096 6

Relict and palimpsest glacial landforms in Nova Scotia, Canada

R.R.Stea
Nova Scotia Department of Natural Resources, Halifax, N.S., Canada

ABSTRACT: The glacial history of Atlantic Canada can be summarized as a struggle between competing ice domains: the local Appalachian ice complex and the Laurentide Ice Sheet centred in Quebec. There is a continuum of landforms present in Nova Scotia, a region dominated by the Appalachian ice complex. Relict landforms are inherited from older glacier advances without substantial modification. Palimpsest landforms are relict forms that have been overprinted by younger ice advances from local centres and bear evidence of the older and younger ice flows. The interplay of the regional ice sheets and local ice caps, and the physiographic variation of Nova Scotia, produced distinct zones where erosional and depositional processes were active, as well as zones where these processes were muted. The development of a major ice divide over Nova Scotia during the Late Wisconsinan produced a corridor characterized by the preservation of relict landforms and palimpsest features.

1 INTRODUCTION

The term palimpsest is derived from the Greek "palimpsestos" which means to scrape again. Palimpsest glacial landforms are a result of the superposition of ice flow events on the landscape and information about previous ice flows can be retrieved from these palimpsest landforms. This paper provides a description of erosional and depositional landforms resulting from several ice flow events and the preservation of older (relict) landforms. Stea and Brown (1989) described "lobate" or "reoriented" drumlins in Nova Scotia, which they attributed to remolding by successive ice flows. This report will update the earlier work of Stea and Brown (1989) and include a detailed

description of a newly-discovered section through a palimpsest drumlin. The author will also describe erosional landforms produced by the action of successive ice flow advances from differing ice centres. Deal (1970) described palimpsest glacial topography as "areas of glaciated topography that owe their basic form to glacial events that pre-date the deposition of the uppermost glacial sediment".

The glacial history of Atlantic Canada can be summarized as a struggle between competing ice domains: the local Appalachian ice complex and the Laurentide Ice Sheet centred in Quebec (Bailey, 1898; MacNeill, 1951; Prest and Grant, 1969) (Fig. 1). These ice sheets crossed Nova Scotia many times in the past. During the

Figure 1 (over). Nova Scotia drumlin fields and the locations of detailed map areas. Area 1 = Figure 14a, Area 2 = Figure 14b, Area 3 = Figure 13, Area 4 = Figure 16; A = Location Figure 4, B = Figure 6a, C = Figure 6b, D = Figure 6c, E = Figure 6d Section, F = Hartlen Point, G = West Lawrencetown; Inset map (after Prest, 1983) showing the Laurentide and Appalachian ice complexes during deglaciation.

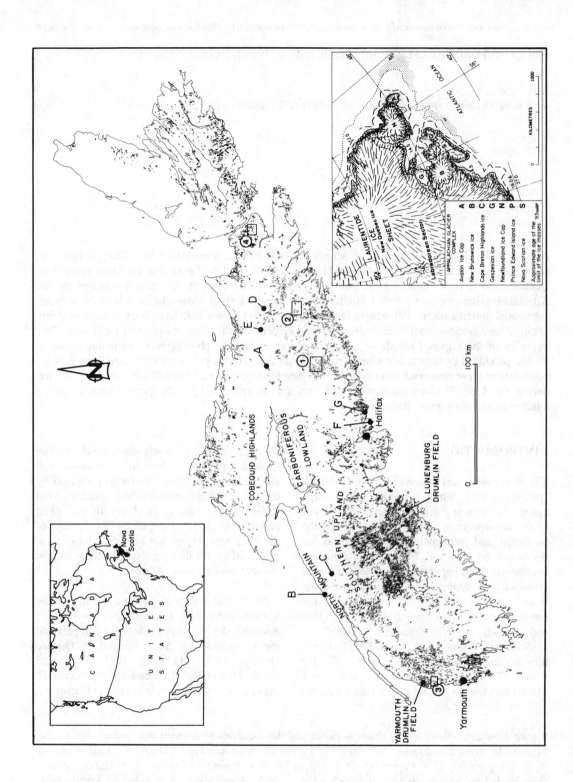

Wisconsin Glaciation the Appalachian ice complex dominated ice flow events in the region. The ice sheets centres shifted from the northwest (New Brunswick) to form local ice divides (Fig. 2). An ice flow sequence has been reconstructed in Nova Scotia through the careful mapping of striae. Ice flow trends can be traced over broad regions of the Province using striae (Stea and Finck, 1984). In addition, crosscutting relations and the preservation of older, weathered striae on lee-side surfaces have allowed for the development of an erosional chronology. The events defined by discrete, regionally mappable trends of striae are termed Ice Flow Phases. The patterns of ice flow mapped by striae are verified by the orientation of glacier landforms such as eskers and drumlins, till fabric and dispersal studies (Stea, 1984). The sequence of Ice Flow Phases has been discerned from superimposed striation sites, and through correlation with stacked till sheets (Stea, 1984). Each of the Ice Flow Phases produced at least one recognizable till sheet, with lodgment and melt-out facies.

Ice flow phases

Phase 1

Striation patterns, distinctive erratics, till fabric, and striated boulder pavements suggest that the earliest and most extensive ice flow in Nova Scotia was eastward then southeastward (Fig. 2). Several widely-spaced striation sites reveal a distinct eastward flow, preserved in lee-side hollows and depressions, later overrun by southeastward-trending striae. In fact, the eastward ice flow may represent a separate, older phase of glaciation. Erratic trains of igneous rocks from the Cobequid Highlands and trains of basaltic rocks from the North Mountain (Fig. 1) are oriented southeastward and southward and can be traced to the Atlantic Coast, up to 120 km down-ice (Grant, 1963; Nielsen, 1976). This phase may represent a continental ice flow. Evidence of the passage of Laurentide ice across New Brunswick, however, is equivocal (Rampton et al., 1984). Anorthosite boulders in western Prince Edward Island, of a presumed Canadian Shield source, suggest that Laurentide ice did cross the region

at some time (Prest and Nielsen, 1987). This flow is presumed to be Early Wisconsinan in age because tills formed during this event overlie interglacial organic beds and several till-forming ice flow events supersede it (Stea et al., in press).

Phase 2

The second major ice flow event was southward and southwestward from the Escuminac Ice Centre in the Prince Edward Island region (Rampton et al., 1984; Fig. 1). This flow phase is analogous to the Acadian Bay Lobe of Goldthwait (1924) and the "Fundian" glacier of Shepard (1930). Goldthwait envisioned southward flow from a Laurentide source across the Gulf of St. Lawrence. Ice flow trends in Prince Edward Island (Prest, 1973) and adjacent New Brunswick (Rampton et al., 1984; Foisy and Prichonnet, 1991) imply a radiating ice centre. The author envisions that the Escuminac Ice Centre, in an early phase, merged with New Brunswick ice (Fig. 1) to produce a more pervasive southward flow. This was followed by more localized ice centres and an ice stream in Chignecto Bay (Fig. 2). This event is recorded by southward striae crossing earlier southeastward-trending striae at many localities on the upland regions of Nova Scotia and New Brunswick. Material from the vast area of redbeds in northern mainland Nova Scotia and Carboniferous basins in the Prince Edward Island region was transported southward onto the metamorphic and igneous rocks of the Cobequid Highlands and Southern Uplands of mainland Nova Scotia. Southward dispersal of distinctive Cobequid Highland erratics occurred with the dispersal of the red material (Grant, 1963). Evidence of southward dispersal of red clastic material from the Bay of Fundy has also been noted in sediment cores in the Gulf of Maine (Schnikter, 1987). This flow event is thought to encompass Middle to Late Wisconsinan time (Stea et al., in press).

Phase 3

During Ice Flow Phase 3 granites from the Southern Upland were transported northward onto the North Mountain basalt cuesta (Hickox,

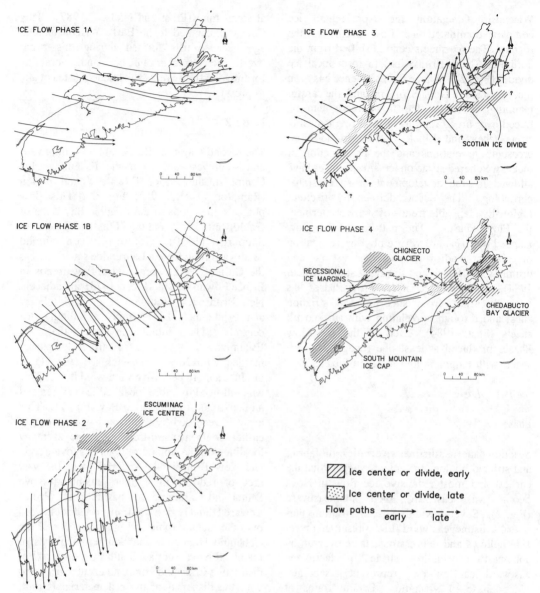

Figure 2. Ice flow lines, centres and phases in Nova Scotia during the Wisconsinan (Cape Breton phases after Grant, 1988). The flow lines were constructed from mapped striations, drumlins, eskers and erratic dispersal trains.

1962). Erratics from the Cobequid Highlands can be found throughout the Carboniferous lowlands to the north (Fig. 3; Stea and Finck, 1984). Northward-trending striae can be traced across the northern mainland of Nova Scotia. This well-documented northward ice flow occurred in response to the development of

an ice divide in southern Nova Scotia termed the Scotian Ice Divide (Fig. 2). This divide may have formed as a result of marine incursion into the Bay of Fundy (Prest and Grant, 1969) or a climatic event (MacNeill and Purdy, 1951; Hickox, 1962). Prolonged ice flow from this divide, across highland regions in northern

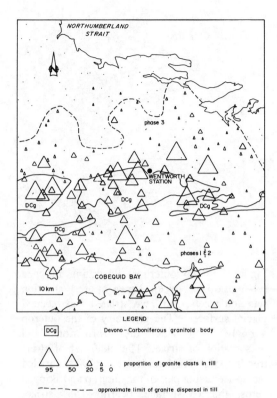

LEGEND

DCg   Devono - Carboniferous granitoid body

△ △ △ ▵ ▵   proportion of granite clasts in till
95 50 20 5 0

— — — — — — approximate limit of granite dispersal in till

Figure 3. Distribution of granitic erratics in the pebble fraction of tills in northern Nova Scotia (after Turner and Stea, 1990).

Nova Scotia that attain 400 m in elevation, suggests buildup of ice locally, rather than simple drawdown. The ice dome off Cape Breton Island proposed by Grant (1977) was probably part of the Scotian Ice Divide. This ice divide developed in the Late Wisconsinan. It formed major moraines on the inner continental shelf that are dated between 14,000 and 15,000 yr B.P. (Stea et al., in press; Stea et al., 1992b).

Phase 4

During this final phase remnant ice caps developed from the Scotian Ice Divide (Fig. 2). Eskers and striae cut across features formed by earlier ice flows (MacNeill in Prest et al., 1972). Ice caps or glaciers that formed over the Chignecto Peninsula and southern Nova Scotia had margins on land marked by moraines, melt-out till, glaciofluvial deposits,

and the pinch-out of till sheets. Ice flow during this last phase was strongly funnelled westward into the Bay of Fundy. Erosional features and deposits relating to these late-glacial ice caps are restricted to low-lying areas.

There is a continuum of landforms present in Nova Scotia. Relict landforms are inherited from older glacier advances without substantial modification. Palimpsest landforms are relict forms that have been overprinted by younger ice advances from local centres and bear evidence of the older and younger ice flows. The interplay of the regional ice sheets and local ice caps, and the physiographic variation of Nova Scotia, produced distinct zones of erosion and deposition (Stea et al., 1989). These zones are a result of the migration of ice divides. Erosional and depositional processes were muted in the areas of the former ice divides. The relict and palimpsest landforms are confined to the regions under the Scotian Ice Divide, which straddles the axis of the Nova Scotia peninsula (Fig. 2).

Erosional relict and palimpsest landforms

Multi-phase striation sets

Many rock surfaces are inscribed with two or more sets of crossing striae and grooves. Surfaces were abraded and plucked by ice flowing in one direction, then reshaped by ice flowing in another. In order to unravel the sequence of events, it is imperative to sort out the sense of ice flow and the crosscutting relationships of facets and striae. The sense of ice flow is determined by stoss/lee relationships and the direction of rat tails or crag and tail forms (Prest, 1983). Rat tails or crag and tail features are elongate ridges found in the lee side of pebbles or other hard obstacles on an otherwise flat surface. The succession of ice flows may be difficult to discern on some flat outcrops. Simple crosscutting relationships are ambiguous and misleading. Older, deeper grooves often appear to crosscut younger, finer striae. Preferential staining of the deeper grooves on some outcrops provides proof of the antiquity of the apparently younger grooves (Fig. 4). The following methods have been

Figure 4. An outcrop in northern Nova Scotia inscribed with two sets of crossing striations (A-Fig. 1). An older, northeastward flow (020°) is found on the lower (foreground) surface. These striae are stained with Fe/Mn oxide. A later, southwestward flow (249°) inscribed a pervasive set of fine striae and grooves on the upper surface.

employed to determine the relative age of striae:

1. The geometry and relief of an outcrop influences the preservation of striae (Fig. 5). Striation sets can be preserved on different faces of the same outcrop without any apparent crosscutting relationships. The older set of striae in Figure 5a trend 155°. The younger ice flow (190°) abraded the part of the outcrop facing up-ice and skipped across the lee surface, bearing 155° striae, creating a bevelled facet. The location of the facet depends on the orientation and relief of the outcrop. Much of the abrasional evidence of the earlier 155° flow event has been erased from the upper surface. If the 155° flow had been later, it would have striated and truncated much of the upper (190°) surface. Older striae are often preserved on the lee face of the outcrop, where abrasion is at a minimum. In some regions the older, striated lee face is stained by Fe and Mn-oxides, a result of periods of ice retreat, weathering and soil formation (Grant, 1989).

2. On many outcrops the evidence of older events has been largely erased from the fresh

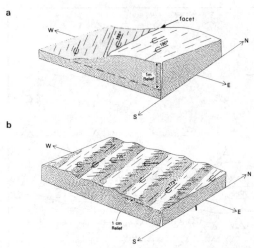

Figure 5. Two outcrops illustrating the age relationships between superimposed striation sets and facets. (a) Two striated and facetted glacial surfaces. The geometry of the outcrop strongly influences the formation and preservation of striae. The older set of striae trends 155°. The later ice flow (190°) abraded the up-ice facing part of the outcrop and skipped across the lee surface, bearing 155° striae, creating a bevelled facet. The location of the facet depends on the orientation and relief of the outcrop. Much of the abrasional evidence of the earlier 155° flow event has been erased from the upper surface. If the 155° flow was later, it would have striated a large part of the upper surface. (b) 205° striae are formed on the up-glacier side of the ridges between grooves bearing an older set of striae trending 173° (modified after Finck and Graves, 1987).

surface by the most recent ice flow. Deeper grooves in the outcrop may preserve the older events, but the younger flow event is represented by fine striae (Lowell et al., 1990). Finer striae tend to be formed on the up-glacier sides of the ridges or grooves and are absent (Fig. 5b) in the intervening hollows, forming again on the adjacent ridges. These lee-side skip zones can be from millimetres to metres across. Boulton (1979) explained these zones as reductions in the effective normal pressure on the debris-rich basal layer. An alternative view is that the striator points (surrounded by ice) lost contact with the bed

(Hallet, 1981). The ability of striators to skip across grooves with less than a centimetre relief lends support to the latter view.

Striae perpendicular to flow have been noted in unidirectional valley glacier settings (Sharp, 1988). These have been explained by secondary flow into lee-side cavities (Sharp, 1988). Some evidence of this can be found on lee surfaces in Nova Scotia where striae can be traced around the stoss surface and into the lee where they curve almost perpendicular to flow. These striae, however, are fresher and generally finer than older sets preserved in the lee face. In many cases the older sets are not restricted to one or several outcrops but can be traced over wide areas of Nova Scotia where outcrops bear similar geometric relationships. Stea et al. (1992a) derived the regional flow lines from their compilation of all striation measurements in Nova Scotia.

Bevelled facets

In northern Nova Scotia ice flow direction changed almost 180° between phases 2 and 3 (Fig. 6). Bedrock surfaces were stossed on one side, then on the opposite side. Distinctive outcrops with bevelled facets were produced (Fig. 6). These are planar surfaces that meet along a distinct line or bevel (Veillette, 1983; Grant, 1989). The surfaces bear striae of opposite sense. The paucity or absence of striae from the latest ice flow on the facet or lee-surface of the earlier flow implies a decoupling of the ice or striators from the bed.

Palimpsest rat tails

Areas south of the Scotian Ice Divide (Fig. 2) were subjected to shifting ice flows in a southward direction. Rat tails are ubiquitous on glaciated slate and metagreywacke surfaces, although they may be barely visible. They form on the lee side of silt to sand-sized metamorphic grains such as garnets. These forms are rapidly worn away when exposed to the elements. On surfaces bearing two or more striation sets, corresponding sets of rat tails can be observed (Fig. 7). Older rat tails are truncated by younger rat tails.

Depositional landforms

Palimpsest and relict drumlins in the West Lawrencetown section

The West Lawrencetown section (Figs. 1, 8) is a wave-eroded part of a large southeast-oriented drumlin several kilometres northeast of Halifax. Bedrock outcrops on the flank of the drumlin. The outcrop reveals an older, parallel set of wide grooves and striae that trend 155° followed by divergent, finer striae trending 180-190° which are inscribed on the north-facing slopes of the larger grooves. The drumlin has a V-shaped form with a main axis trending 150° and a minor axis trending 185° (Fig. 8).

The southern, tapered end of the drumlin has been truncated by the sea. Exposures show two compositionally-distinct till units termed the Hartlen and Lawrencetown tills (Fig. 8). The lower, Hartlen Till, was first defined by Stea and Fowler (1979) and the upper, Lawrencetown Till, by Grant (in Williams et al., 1985). The type section for the Hartlen Till is at Hartlen Point a few kilometres east of West Lawrencetown (Fig. 1). The Lawrencetown Till type section is designated as Sandwich Point, south of Halifax (Grant in Williams et al., 1985). The Lawrencetown Till can be divided into an upper, sandier, and a lower, clay-rich, facies. It is thickest in the south-trending axis of the drumlin, whereas the Hartlen Till is thickest in the southeastern-trending axis. It is clear that the southeastern axis is made up largely of the lower Hartlen Till while the southern axis is a contructional feature formed primarily of the upper Lawrencetown Till.

The Hartlen Till is greyish, silty and very compact. It contains abundant metagreywacke boulders (>80%) of the underlying Cambro-Ordovician Meguma Group. Many bullet-shaped, stoss-lee boulders are embedded in the Hartlen Till (Fig. 9). The trend of the long axes of these large boulders reflects the direction of ice flow that emplaced the till (Hicock and Dreimanis, 1985). The shape indicates lodgment or subsequent moulding by overriding ice (Kruger, 1984; Clark and Hansel,

Figure 6. Examples of facetted glacial surfaces in northern mainland Nova Scotia. (a) This basaltic outcrop records two opposing ice flows. The earlier flow (Phase 1, Fig. 2) is parallel to pencil - 125°, the later flow (Phase 3, Fig. 2) is parallel to the compass - 310°. The facets meet along a distinct line (see Fig. 1 for location marked as B). (b) Granitic outcrop with early ice flow (Phase 1) at 120° (closest compass), followed by a flow to 320° (Location C - Fig. 1). (c) Early southwest ice flow (stained surface) followed by flow to north (9°) indicated by straie on the upper surface (Location D - Fig. 1). (d) Early flow (this time on the upper surface and marked by compass) toward 142°, followed by flow toward 345° (Location E - Fig. 1).

1989). The long axes of boulders in the Hartlen Till (Fig. 10) have a dominant southeastward trend, as do striae on their upper surfaces (Fig. 10). These trends are concomitant with the older set of grooves and striae on the bedrock surface adjacent to the drumlin and the main drumlin axis. These stoss-lee boulder surfaces were inscribed with a second set of striae, trending 180-190°. Some of the boulders were rotated parallel to the second

bedrock striation set and the southward drumlin axis. The Hartlen Till has a strong, unidirectional clast fabric with an eigenvector of 155°, parallel to the older set of striae in the rock core (Fig. 11).

A reddish, muddy till unit (Lawrencetown Till) overlies the Hartlen Till with a knife-sharp contact. It is distinguished from the Hartlen Till by its red colour and abundant erratic

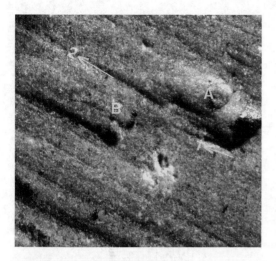

Figure 7. Palimpsest rat tails on a metagrey-wacke surface. Earlier forms (A) are parallel to a relict striation set (1) that has been nearly erased from the surface but remains in the deeper grooves (arrow). A later flow produced the main set of southeastward-trending striations and another set of parallel rat tails (C). The largest grain is 0.5 mm across. The inferred ice flows are shown by arrows.

content. Distinctive hornblende-bearing syenogranite, foliated granodiorite, and volcanic erratics from the Lawrencetown Till have source areas in the Cobequid Highlands, 100 km north of the section (Fig. 12). Fine-grained micrographic syenogranite erratics and foliated syenogranite and granodiorite within the Lawrencetown Till are thought to have the most restricted ranges of the erratics. Their source areas (source areas C, D, and G, Fig. 12) are due north of the section, confined between 357° and 010°. Entrainment of red mudstone from areas north of the Cobequids or in the Minas Basin area produced the red matrix of the Lawrencetown Till.

Stoss-lee boulders in the Lawrencetown Till (Fig. 10) have a wider range than those in the Hartlen Till, exhibiting many boulders with the same orientation as the Hartlen Till. There is, however, a distinct southward mode in long-axes orientations. Striation trends on their surfaces are divergent, with a peak parallel to

the younger striation set on the bedrock outcrop and the south axis of the drumlin. Most long axes of boulders and surface striae are oriented parallel to the Hartlen Till fabric. Till fabric in the Lawrencetown Till has an eigenvector of 176° (Fig. 11). The abundance of southeast-ward-trending boulders in the Lawrencetown Till is paradoxical. The southeastward-trending boulders in the Lawrencetown Till may have been inherited from the Hartlen Till, largely in their original orientation.

The West Lawrencetown drumlin has well-defined stoss and lee form. The stoss end faces northwest while the lee side gently slopes toward 145° (Fig. 8). The Hartlen Till comprises most of Section B (Fig. 8) on the lee side of this drumlin. Section A is located on a lobe-like extension of the drumlin oriented 193° (arrow, Fig. 8). In this section, Lawrencetown Till forms much of the drumlin's constructional topography. This suggests that the lobe-like extension of the drumlin, dominated by Lawrencetown Till, was accreted upon the earlier, tapered, southeast-trending drumlin formed of Hartlen Till.

Regional evidence for relict and palimpsest drumlins

The internal structure of Nova Scotia drumlins generally resembles a layer cake of till sheets. The most common situation is a grey core till (Hartlen Till) formed by southeastward ice flow across grey metasedimentary bedrock and a distinctly red mantle till formed by southward to southwestward ice flow (Lawrencetown Till). Drumlins in the Yarmouth field of southern Nova Scotia exhibit the greatest number of stacked units, capped by a coarse-textured till that represents deposition by a late-stage ice mass centred inland (Ice Flow Phase 3; Figs. 2, 13). In Yarmouth, drumlins exhibit as many as 13 lithostratigraphic units in superposition. They are clearly erosional, as most of the units are subhorizontal and truncated by the drumlin form. Till units within the Yarmouth field represent all four ice flow phases in Nova Scotia. An interglacial sand bed is interbedded with till units within the drumlin. Most of the drumlins are relict forms

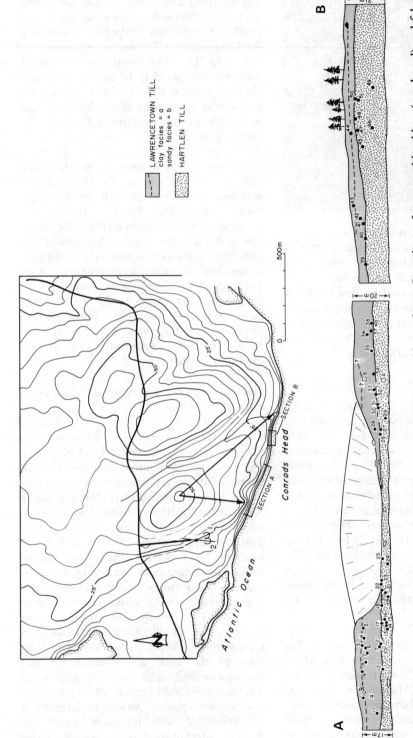

Figure 8. Map of the drumlins at West Lawrencetown and stratigraphic sections. Locations of measured boulders (numbered) and fabric locations. Arrows depict the primary and secondary axes of the palimpsest drumlin. The striation site on the flank of the drumlin records two ice flows 155° and 180-190°. These striae are parallel to the drumlin axes.

LAWRENCETOWN TILL
clay facies = a
sandy facies = b

HARTLEN TILL

150

Figure 9. A striated boulder with a bullet shape, indicating transport by sliding in the basal zone of traction (after Stea and Finck, 1984).

dominated by the core Hartlen Till and oriented southeast. At least two ice advances have succeeded the deposition of the Hartlen Till. These ice advances produced V-shaped, palimpsest drumlins in many fields (Fig. 14). In the Yarmouth field (Fig. 13), reworking by successive ice flows (Phases 2 and 3; Fig. 2) has rendered the original form unrecognizeable and produced forms of many shapes.

Figure 15 shows the relationship between form and stratigraphy that differentiates relict and palimpsest forms. Nucleating agents for initiation of the relict drumlins (phase 1; Fig. 2) may have been pre-existing tills or waterlain deposits. Palimpsest drumlins nucleated from pre-existing southeastward-trending forms. Flow from the local Nova Scotia ice caps (Phases 3, 4; Fig. 2) also formed some drumlins. These drumlins can have irregular shapes, but some are oriented with their stoss end facing inland.

Palimpsest moraines

Ice flow from the Scotian Ice Divide (Fig. 2) was strongly funnelled through the submarine channels in the Gulf of St. Lawrence. This vigorous ice flow produced palimpsest forms

resembling rogen or ribbed moraine. Four of these features are shown on Figure 16. Like rogen moraine at the type area (Lundqvist, 1969) these forms grade down-ice (northward) into northward-trending drumlins. They are composed of a reddish, muddy till, similar to the Lawrencetown Till described earlier. South of these features only southeastward-trending drumlins are found. The northernmost moraine is composed of a southeastward-trending ridge, hooked northward at the tips (1, 2; Fig. 16). Southward-facing stoss ends are developed at the south end of the northward-trending hooks. There is a clear separation between the southeastward ridge and the northward-trending drumlin (Fig. 16). This may be a result of subsequent stream erosion or may reflect the separation of the northward drumlin from the relict southeastward ridge. Coastal exposures of these ridges are composed of a lodgment till formed by a Wisconsinan southeastward ice flow (Ice Flow Phase 1) capped by another lodgment till formed by northwestard ice flow (Ice Flow Phase 3; Stea and Mott, 1990). Large eastward-trending ridges found in Cape Breton Island (Grant, 1988) are also interpreted as relict features.

2 DISCUSSION

Rat tails are generally considered to have been formed by ice erosion (e.g. Prest, 1983). Recently an alternative hypothesis has been put forward by Shaw and Sharpe (1987) who claim that the features are formed by subglacial water. They invoke turbulent flow separation around obstacles. The case for a subglacial-fluvial origin rests on the analogy of similar features in soft sediment (flutes) and reproduction of these in flume tanks. Because the effects of erosion at the base of a glacier are difficult to observe directly, it is impossible to verify the formation of these features by glacier ice. Parallel striae, however, have no analogue in fluvial systems. Slickenside surfaces may provide a reasonable analogue to the formation of rat tails. These bear both striae and rat tails (Means, 1987). On moving fault planes, rat tails or trails are formed when areas are

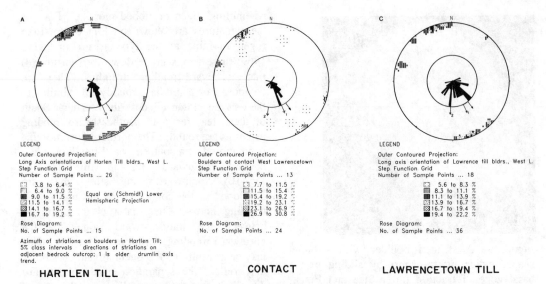

HARTLEN TILL   CONTACT   LAWRENCETOWN TILL

Figure 10. Lower hemispheric, equal area stereographic plot of the trend and plunge of the long axis of stoss-lee boulders at West Lawrencetown. Inner rose diagram indicates the azimuth of upper boulder surface striation trends. (A) Hartlen Till, (B) Boulders at the contact, (C) Lawrencetown Till. The drumlin trend is indicated by the arrow. The striation symbols are the trends of bedrock striations from Figure 8.

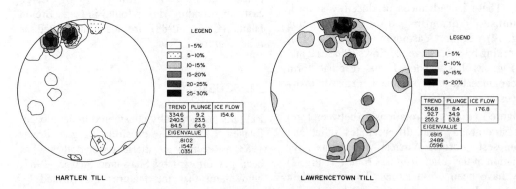

Figure 11. Lower hemispheric, equal area projections of the plunge of the A-axis of elongate (3:1) stones in the Hartlen (A) and Lawrencetown tills (B). Patterns represent percentages of 50 points in 1% area of projection. The location of these fabrics is indicated on Figure 8 (A, B). The minor southwest-trending population in the Lawrencetown Till is from boulders 41-43 (Fig. 8). These may reflect the late southwestward ice flow during Phase 4 (Fig. 2).

sheltered from abrasion by fault gouge by resistant asperities (Means, 1987). The simplest hypothesis to explain the association of rat tails and striae on glaciated surfaces would be abrasion by debris-rich ice. This argument is strengthened when palimpsest rat tails are found with multi-phase striation sets.

Palimpsest erosional forms such as bevelled facets can be mistaken for forms attributed to water erosion such as sichelwannen (Shaw and Sharpe, 1987). In many regions of Nova Scotia, older facets preserved in lee surfaces are devoid of striae. This may be the result of a previous cycle of weathering. These curved,

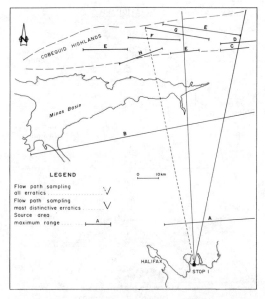

Figure 12. Source areas of the boulders in the West Lawrencetown Section (Lawrencetown Till). Many of these erratics came from the Cobequid Highlands, an upland region attaining 300 m in height. Erratic assemblages: (A) granodiorite-monzogranite-tonalite, (B) red sandstone-siltstone-conglomerate-limestone-gypsum, (C) foliated granodiorite, (D) quartzite, (E) volcanics, (F) diorite, (G) syenogranite-granite-hornblende granite.

smoothed surfaces can resemble sichelwannen.

Time-independent glacial and fluvial process models have dominated much of the present thinking on drumlins and other landforms. Form, and stratigraphic and sedimentologic variability have been explained by process models with the tacit assumption that drumlins are formed by the latest ice advance. Nova Scotia, however, has many landforms that are relict from earlier glaciations. In fact, there is a progression from relict to palimpsest glacial forms. In other regions the effects of earlier glaciations on drumlin morphology are being recognized (Krall, 1977; Coude, 1989; Rouk and Raukas, 1989). Lundqvist (1989) now interprets rogen moraine as a palimpsest feature. He believes that rogen moraine in the type area was formed

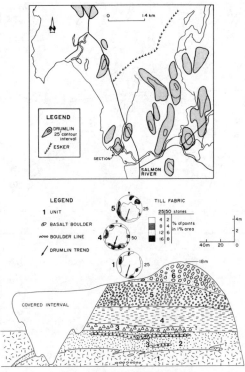

Figure 13. Drumlin map and section in the Yarmouth drumlin field. Note the irregular shapes of the drumlins and the esker crossing the terrain fabric. Till Unit 1 - Red Head Till -- formed during Phase 1. Unit 2 - Salmon River Sand (Sangamon); Till Units 3-4-5 (Saulnierville Till) formed during Phase 2; Till Unit 6 (Beaver River Till) - Phases 3 or 4 (after Stea and Brown, 1989).

in two or more steps by remolding of pre-existing forms.

The landscape of Nova Scotia has been smeared by glaciations of decreasing vigour or intensity during the Wisconsinan glacial stage (Grant, 1989). The last ice caps left little evidence of their passage. Notable exceptions are relict drumlins in eastern Nova Scotia that were overridden by ice of late-glacial age (Stea and Mott, 1989). Till deposits of the last phases of Late Wisconsinan ice flow (Phases 3 and 4; Fig. 3) are generally thin and patchy. The preservation of relict forms can be attributed to the development of local ice centres

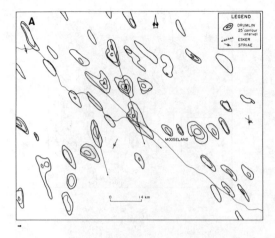

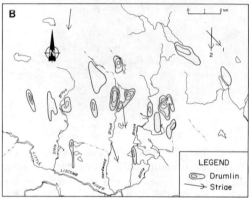

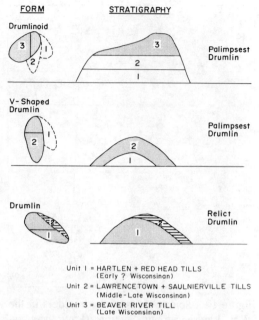

Unit I = HARTLEN + RED HEAD TILLS
(Early ? Wisconsinan)
Unit 2 = LAWRENCETOWN + SAULNIERVILLE TILLS
(Middle-Late Wisconsinan)
Unit 3 = BEAVER RIVER TILL
(Late Wisconsinan)

Figure 15. Relationship between form and stratigraphy in Nova Scotia drumlins. The diagrams show the plan view and cross-section of typical relict and palimpsest drumlins in Nova Scotia. Phase 1 (Ice Flow Phase 1; Fig. 3) drumlins are relict forms composed mainly of tills believed to be Early Wisconsinan in age. Cross hatching represents passive deposition of Phase 2 tills (e.g. Lawrencetown Till) on the drumlin without form modification. Phase 2 and 3 drumlins are palimpsest forms. The dashed lines represent precursor forms. The solid line through the drumlin plan view indicates the line of cross-section represented on the right.

Figure 14. (a) Drumlin map of the Moose River drumlin field near Mooseland, Nova Scotia (Location on Fig. 1). Note V-shaped palimpsest drumlin forms and bedrock striae parallel to the drumlin primary and secondary axes. The long arrows denote the inferred flow paths from the palimpsest forms. (b) Drumlin map of the Liscomb drumlin field.

and divides over older glacial topography. These zones are generally cold-based (Denton and Hughes, 1981) and are sites of low velocity and muted erosion. Evidence of palimpsest and possible relict till plains under the New Quebec ice divide has been documented by Richard *et al.* (1991) who found that the surface till was characterized by abundant interstadial or interglacial pollen.

3 CONCLUSIONS

The history of glaciations in a region such as Maritime Canada on the eastern margin of the Laurentide/Appalachian ice complex can be unraveled by the careful mapping of striae, grooves, rat tails or small scale crag and tail features and abraded surfaces. In order to unravel the sequence of events, it is imperative to sort out the sense of ice flow and the crosscutting relationships of facets and striae. The succession of ice flows may be

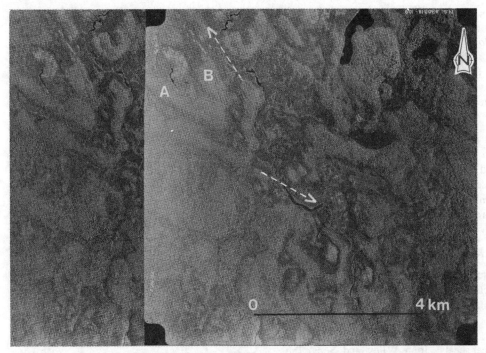

Figure 16. Stereo pair of palimpsest "ribbed moraine" in northern Nova Scotia (see Fig. 1), composed of (A) relict Phase 1 drumlin and (B) palimpsest Phase 3 drumlin. Regional ice flow events are shown by arrows. Note other crescentic forms within field of palimpsest moraines and more irregular features. All are formed of reddish, muddy till.

difficult to discern on some flat outcrops. Simple crosscutting relationships are ambiguous and misleading. These methods were employed to determine the relative age of striae:

1. The relief and facing direction of an outcrop influences the preservation of striae. Striation sets are preserved on different faces of the same outcrop without any apparent crosscutting relationships. Older striae are often preserved on the lee face of the outcrop, where abrasion is at a minimum. Sometimes the older, striated lee surface is recognized by staining, a result of periods of ice retreat, weathering and soil formation. If the shift in ice flow is a result of ice dynamics rather than ice buildup and retreat, the older flow can be determined by stoss-truncation.

2. On many outcrops the older events have been largely erased from the fresh surface by the most recent ice flow, except in deeper grooves and ridges. The younger flow event is represented by finer striae. The finer striae tend to be formed on the up-glacier sides of the ridges or grooves.

Palimpsest glacial landforms are defined as forms whose structure is a result of the superposition of ice flow events on the landscape. A continuum of landforms is recognized in Nova Scotia, including relict and palimpsest glacial erosional and depositional forms. The preservation of relict forms in Nova Scotia is a result of the development of ice divides over older glacial topography.

4 REFERENCES

Bailey, L. W. 1898. Report on the geology of southwestern Nova Scotia. *Geological Survey of Canada, Annual Report* 9, pt. M: 30-60.

Boulton, G. S. 1979. Processes of glacier erosion on different substrata. *Journal of Glaciology* 23, no. 89: 15-38.

Clark, P. U. & A. K. Hansel 1989. Clast ploughing, lodgement and glacial sliding over a soft glacier bed. *Boreas* 18: 201-207.

Coude, A. 1989. Comparative study of three drumlin fields in western Ireland: geomorphic data and genetic implication. *In* J. Menzies and J. Rose (editors), Subglacial Landforms-Drumlins, Rogen Moraine and Associated Subglacial Landforms. *Sedimentary Geology* 62: 321-335.

Deal, D. E. 1970. Thin till sheets and palimpsest glacial topography; *In* Program and Abstracts, American Quaternary Association, 1st annual meeting, Seattle, Washington, p. 33.

Denton, G. H. & T. J. Hughes 1981. The last great ice sheets; John Wiley and Sons, Inc., Toronto, Ontario, 484 p.

Finck, P. W. & R. M. Graves 1987. Glacial Geology of Halifax and Sambro (NTS sheets 11D/12 and 11D/05), Nova Scotia; Nova Scotia Department of Mines and Energy, Map 87-2, scale 1:50 000.

Foisy, M. & G. Prichonnet 1991. Reconstruction of glacial events in southeastern New Brunswick. *Canadian Journal of Earth Sciences* 28: 1594-1612.

Goldthwait, J. W. 1924. Physiography of Nova Scotia. *Geological Survey of Canada, Memoir 140*, 179 p.

Grant, D. R. 1963. Pebble lithology of the tills of southeast Nova Scotia; unpublished M.Sc. thesis, Dalhousie University, Halifax, Nova Scotia, 235 p.

Grant, D. R. 1977. Glacial style and ice limits, the Quaternary stratigraphic record, and changes of land and ocean level in the Atlantic Provinces, Canada. *Geographie Physique et Quaternaire* xxxi, no. 3-4: 247-260.

Grant, D. R. 1988. Surficial geology, Cape Breton Island, Nova Scotia; Geological Survey of Canada, Map 1631A, scale 1:125 000.

Grant, D. R. 1989. Quaternary geology of the Atlantic Appalachian region of Canada; Chapter 5 in Quaternary Geology of Canada and Adjacent Greenland, R. J. Fulton (ed.); Geological Survey of Canada, Geology of Canada no. 1 (also Geological Society of America, The Geology of North America, v. k-1), p. 393-440.

Hallet, B. 1981. Glacial abrasion and sliding: their dependance on the debris concentration in basal ice. *Annals of Glaciology* 2: 23-28.

Hicock, S. R. & A. Dreimanis 1985. Glacio-tectonic structures as useful ice-movement indicators in glacial deposits: four Canadian case studies. *Canadian Journal of Earth Sciences* 22: 339-346.

Hickox, C. F., Jr. 1962. Pleistocene geology of the central Annapolis Valley, Nova Scotia; Nova Scotia Department of Mines, Memoir 5, 36 p.

Krall, D. B. 1977. Late Wisconsinan ice recession in East Central New York. *Geological Society of America Bulletin* 88: 1697-1710.

Kruger, J. 1984. Clasts with stoss-lee form in lodgement tills: a discussion. *Journal of Glaciology* 30: 241-243.

Lowell, T. V., J. S. Kite, P. E. Calkin & E. F. Halter 1990. Analysis of small scale erosional data and a sequence of late Pleistocene flow reversal, northern New England. *Geological Society of America Bulletin* 102: 74-85.

Lundqvist, J. 1969. Problems of the so-called Rogen Moraine; Sverige Geol. Unders. Ser. no. 618, 32 p.

Lundqvist, J. 1989. Rogen (ribbed moraine) - identification and possible origin; *In* J. Menzies and J. Rose (editors), Subglacial Landforms-Drumlins, Rogen Moraine and Associated

Subglacial Landforms. *Sedimentary Geology* 62: 281-292.

MacNeill, R. H. 1951. Pleistocene geology of the Wolfville area, Nova Scotia; M.Sc. thesis, Acadia University, Wolfville, Nova Scotia; Nova Scotia Department of Mines and Energy, Thesis 241, 59 p.

MacNeill, R. H. & C. A. Purdy 1951. A local glacier in the Annapolis-Cornwallis Valley (abstract). *Proceedings of the Nova Scotian Institute of Science* 23, pt. 1: 111.

Means, W. D. 1987. A newly recognized type of slickenside striation. *Journal of Structural Geology* 9: 585-590.

Nielsen, E. 1976. The composition and origin of Wisconsinan tills in mainland Nova Scotia; Ph.D. thesis, Dalhousie University, Halifax, Nova Scotia, 256 p.

Prest, V. K. 1973. Surficial deposits of Prince Edward Island; Geological Survey of Canada, Map 1366A.

Prest, V. K. 1983. Canada's heritage of glacial features; Geological Survey of Canada, Miscellaneous Report 28, 119 p.

Prest, V. K., D. R. Grant, R. H. MacNeill, I. A. Brooks, H. W. Borns, J. G. Ogden, III, J. F. Jones, C. L. Lin, T. W. Hennigar & M. L. Parsons 1972. Quaternary geology, geomorphology and hydrogeology of the Atlantic Provinces; 24th International Geological Congress, Excursion Guidebook, A61-C61, 79 p.

Prest, V. K. & D. R. Grant 1969. Retreat of the last ice sheet from the Maritime Provinces - Gulf of St. Lawrence region; Geological Survey of Canada, Paper 69-33, 15 p.

Prest, V. K. & E. Nielsen 1987. The Laurentide ice sheet and long distance transport; *In* Kujansuu, R. and Saarnisto, M. (editors),

INQUA Till Symposium, Finland 1985; Geological Survey of Finland, Special Paper 3, p. 91-102.

Rampton, V. N., R. C. Gauthier, J. Thibault & A. A. Seaman 1984. Quaternary geology of New Brunswick. *Geological Survey of Canada, Memoir* 416: 77 p.

Richard, P. J. H., M. A. Bouchard & P. Gangloff 1991. The significance of pollen-rich inorganic lake sediments in the Cratere du Nouveau-Quebec area, Ungava, Canada. *Boreas* 20: 135-149.

Rouk, A.-M. & A. Raukas 1989. Drumlins of Estonia; *In* J. Menzies and J. Rose (editors), Subglacial Landforms-Drumlins, Rogen Moraine and Associated Subglacial Landforms. *Sedimentary Geology* 62: 371-384.

Schnikter, D. 1987. Late Glacial paleoceanography of the Gulf of Maine; abstract, International Union of Quaternary Research, 12 International Congress, Programme and Abstracts, p. 260.

Sharp, R. P. 1988. Living Ice: Understanding glaciers and glaciation; Cambridge University Press, 225 p.

Shaw, J. & D. R. Sharpe 1987. Drumlin formation by subglacial meltwater sedimentation. *Canadian Journal of Earth Sciences* 24: 2316-2322.

Shepard, F. P. 1930. Fundian faults or Fundian glaciers. *Geological Society of America Bulletin* 41: 59-674.

Stea, R. R. 1984. The sequence of glacier movements in northern mainland Nova Scotia determined through mapping and till provenance studies; *In* Correlation of Quaternary Chronologies, ed. W. C. Mahaney, p. 279-297.

Stea, R. R. & Y. Brown 1989. Variation in drumlin orientation, form and stratigraphy relating to successive ice flows in southern and central Nova Scotia. *Sedimentary Geology* 62: 223-240.

Stea, R. R., H. Conley & Y. Brown 1992a. Surficial Geology of the Province of Nova Scotia; Nova Scotia Department of Natural

Resources, Mines and Energy Branches Map 92-3, scale 1:500 000.

Stea R. R., G. B. J. Fader & R. Boyd 1992b. Quaternary seismic stratigraphy of the inner shelf region, Eastern Shore, Nova Scotia; *In* Current Researcxh, Part D; Geological Survey of Canada, Paper 92-1D, p. 179-188.

Stea, R. R. & P. W. Finck 1984. Patterns of glacier movement in Cumberland, Colchester, Hants and Pictou Counties, northern Nova Scotia; *In* Current Research, Part A; Geological Survey of Canada, Paper 84-1A, p. 477-484.

Stea, R. R. & J. H. Fowler 1979. Minor and trace-element variations in Wisconsinan tills, Eastern Shore region, Nova Scotia; Nova Scotia Department of Mines and Energy, Paper 79-4, 30 p.

Stea R. R. & R. J. Mott 1989. Deglaciation environments and evidence for glaciers of Younger Dryas age in Nova Scotia, Canada. *Boreas* 18: 169-187.

Stea, R. R. & R. J. Mott 1990. Quaternary Geology of Nova Scotia: Guidebook for 53rd Annual Friends of the Pleistocene Field Excursion; Nova Scotia Department of Mines and Energy, Open File Report 90-008, 85 p.

Stea, R. R., R. G. Turner, P. W. Finck & R. M. Graves 1989. Glacial dispersal in Nova Scotia: a zonal concept; *In* Drift Prospecting, ed. R. N. W. Dilabio and W. B. Coker; Geological Survey of Canada, Paper 89-20, p. 155-169.

Stea R. R., R. J. Mott, D. F. Belknap & U. Radtke in press. The Pre-Late Wisconsinan Chronology of Nova Scotia, Canada; *In* The Last Interglaciation/Glaciation Transition in North America, eds. P. U. Clark and P. D. Lea Geological Society of America, Special Paper.

Turner, R. G. & R. R. Stea 1990. Intepretation of till geochemical data in Nova Scotia, Canada, using mapped till units, multi-element anomaly patterns and the relationship of till clast geology

to matrix geochemistry. *Journal of Geochemical Exploration* 37: 225-254.

Veillette, J. J. 1983. Les polis glaciaires au Temiscamingue: une chronologie relative; dans Recherches en Cours, Partie A; Commission geologique du Canada, Etude 83-1A, p. 187-196.

Williams, G. L., L. R. Fyffe, R. J. Wardle, S. P. Colman-Sadd, R. C. Boehner & J. A. Watt 1985. Lexicon of Canadian Stratigraphy, Volume VI, Alantic Region; Canadian Society of Petroleum Geologists, 572 p.

Formation and Deformation of Glacial Deposits, Warren & Croot (eds) © 1994 Balkema, Rotterdam

Deglaciation during a transgression – Late glacial depositional er in the Lund area, SW Sweden

Kärstin Malmberg Persson & Erik Lagerlund
Department of Quaternary Geology, Lund University, Sweden

ABSTRACT: Sections west of the city of Lund, SW Sweden, display a stratigraphical sequence where a lodgement till is overlain by melt-out and flow tills and glaciofluvial sand and gravel. These glacial deposits are capped by a glacioaquatic sediment consisting of silt, clay and diamicton. The environmental development that can be reconstructed from the sedimentological record starts with subglacial deposition below active ice coming from the east, which is the last recorded active ice movement direction in the area. The ice stagnated and had partly disintegrated when the area was inundated and a glacioaquatic sediment was deposited simultaneously with the final melting of the buried ice. This glacioaquatic sediment is a part of a regional lithostratigrafic unit, the Lund Diamicton, which has been identified in the Öresund area.

1 INTRODUCTION

This paper describes the depositional environments during the Weichselian final deglaciation in a part of western Skåne, at the city of Lund.

It has been demonstrated that the retreat of the active ice took place subaerially, with a relative sea level lower than 10-15 m above present sea level in the Öresund area (Lagerlund 1980, 1987a; Adrielsson 1984). After deglaciation, the ground surface was exposed to periglacial processes and ice wedge polygons and ventifacts developed (Berglund & Lagerlund 1981; Lagerlund 1987b; Malmberg Persson & Lagerlund 1990). This surface was inundated during a subsequent transgression and a clay bed of varying thickness drapes the wind-polished clasts, which are still lying in situ. The clay is overlain by a glaciolacustrine sediment sequence consisting of diamicton, clay and silt. This stratigraphic unit was named the Lund Diamicton and was described by Malmberg Persson & Lagerlund (1990).

The Värpinge locality (Fig. 1) discussed in this paper, bears evidence that the transgression and deposition of glacioaquatic sediments took place simultaneously with the final melting of stagnant ice.

The section at Värpinge was displayed during road construction. A total of 400 m of sections was studied (Fig. 2). The area lies c. 15 m above sea level and has a low relief. A brief description of the site was given in Lagerlund & Malmberg Persson (1990).

A representative part of the section was lithologically mapped in detail (Fig. 3) and clast fabric was studied at several places. 25 particles, more than 2 cm long and with an a:b axis ratio of at least 1.5, were measured in each analysis. Statistical treatment was made according to the eigenvalue method (Mark 1973). The petrographical composition of the 3 - 8 mm gravel fraction was studied by means of a binocular microscope.

2 DESCRIPTIONS OF SEDIMENTS AND INTERPRETATIONS OF GENESIS

The sediments are divided into three units: A, B and C (Fig 2). Units A and B are mainly diamictons, while the uppermost unit C is complex.

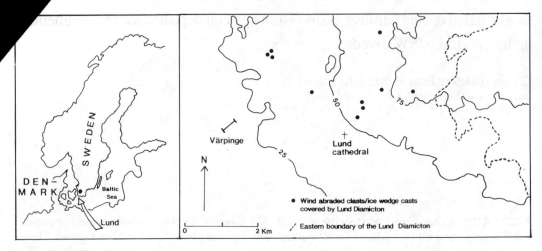

Fig. 1. Location map

2.1 Unit A, description

Unit A is a clayey, sandy, massive diamicton, very hard and dense and with a uniform grain-size distribution. Up to four meters of this unit was displayed in the sections and according to geotechnical investigations it continues more than two meters downwards. It has a fissility caused by subhorizontal and vertical, discontinuous joints which split the sediment into diamond-shaped pieces, 2-3 cm thick, but not associated with slickensides or visible shearing. Flat-iron shaped clasts are common, often with parallel striae on the upper surfaces.

Clast fabric was investigated in 5 places in unit A (Fig. 2 and Table 1). Generally, fabric is characterized by moderate to strong preferred orientations with an easterly direction of the main vector (with the exception of analysis 15). In one analysis (16), very steeply dipping clasts were recorded (plunge of V_1: 56°), but a rather strong clustering of a-axes in an easterly direction was retained. At this place, the fissility pattern was not subhorizontal, but inclined parallel to the clast orientation.

Unit A has a "NE-type" petrographical composition dominated by rock types derived from the central and NE parts of Skåne, mainly Silurian shale and crystalline rocks.

2.2 Unit A, interpretation

Unit A is interpreted as a lodgement till on basis of the dense character, the uniform, massive texture, the striated flat-iron shaped cobbles and the consistent clast fabric (e.g. Dreimanis 1989). The preferred orientation of the clasts is interpreted to reflect a last active ice movement from the east in the area. The fissility is likely related to stress applied by the moving glacier. The steeply dipping clasts and fissility at fabric analysis site 16 are interpreted as a result of glaciotectonic deformation within unit A, although the massive character of the unit cannot reveal any visible deformation structures.

2.3 Unit B lower part, description

Unit B overlies unit A with a gradational contact. The lower part of unit B is a clayey, sandy, moderately dense diamicton which contains subhorizontal, discontinuous laminae of massive or laminated medium sand, up to 1 cm thick. It has a fissility pattern consisting of undulating, subhorizontal joints with 1-2 cm spacing. It is however markedly less dense than unit A.

A 20-30 cm thick horizontal bed of clast-supported sandy gravel with interbeds of silty matrix-supported diamicton is present in the lower part of unit B. This bed has an erosive lower contact, but the upper contact is non-erosive.

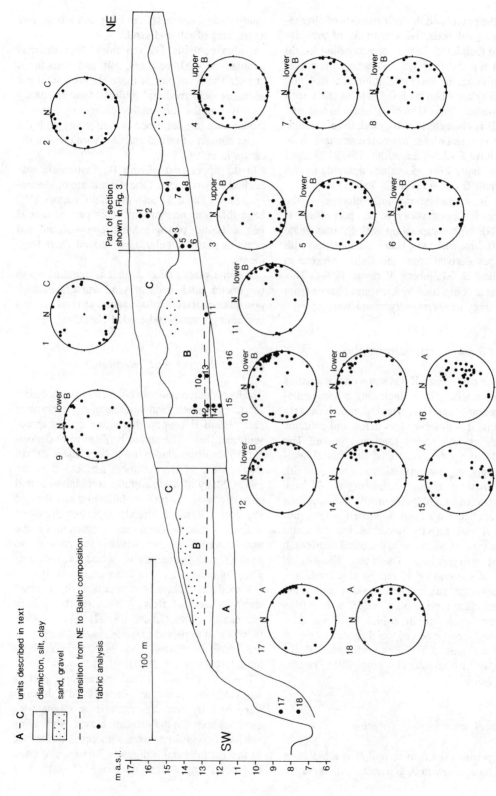

Fig. 2. The section at Värpinge subdivided into three units. Clast fabrics are presented in lower hemisphere Schmidt projections. Statistical parameters are shown in Table 1.

Above the gravel bed the unit consists of alternating horizontal beds, 1-2 cm thick, of gravelly sand and diamicton. In the corresponding lateral position at c. 2-4 m (Figs. 3,4), where the gravel bed is missing, the horizontal joints that make up the fissility are filled with 0.1 - 1 cm thick well sorted medium sand. The clast fabric in this part of unit B is characterized by random to moderately strong orientations, where the stronger main vectors have E - ENE azimuths. There is a trend of successively lower S_1-values upwards in this part of unit B (analyses 5-13, Table 1).

There is a gradual upwards change in the petrographical composition of this part of unit B from a NE-type composition with Silurian shale and crystalline rocks to a Baltic composition with rock types derived from the Baltic depression (Lagerlund & Malmberg Persson 1990). This change is accompanied by a textural change from clayey-sandy to clayey-silty diamicton.

2.4 Unit B lower part, interpretation

The lower part of unit B has some characteristics in common with unit A, including a homogeneous appearance and clast orientation with easterly directions. It is however less dense and contains interbeds of well sorted sand and gravel. The depositional upper contacts of the fluvial sediments suggest an interpretation as a melt-out till, as deformation, erosion or shearing would have been expected in a lodgement till. The interbeds are thought to have been deposited where subglacial fluvial activity occurred intermittently with melt-out of debris from the basal debris-rich parts of stagnant ice. The more fine-bladed fissility in this unit could represent unloading.

The upwards successively weaker and more scattered clast fabrics may reflect a transition from basal melt-out till, which is expected to have a well-preserved englacial fabric, to supraglacial melt-out till. The significance of the change in petrographical composition is discussed below.

2.5 Unit B upper part, description

The uppermost 0-1.5 m of unit B is a heterogeneous, loose diamicton, generally with a clayey,

sandy composition. It contains intraclasts and aggregates of silt and sand.

It also contains 1-2 cm thick discontinuous laminae of massive sand, silt and sometimes gravel. The main part of these are horizontal but some are deformed and folded. Many are partly dissolved in the surrounding diamicton. Some of the laminae are deformed in conformity with the upper contact of unit B and also with the stratification in unit C.

In the upper part of unit B, clast fabric was studied at two places. Orientations are moderately strong (Table 1) and the main vectors (V_1) have different orientations. This part of unit B has a Baltic petrographical composition and contains many angular, sharp-edged chert fragments.

In some places (Fig. 2) unit B contains up to two meter thick beds of sand and gravel with postsedimenatary deformation structures and containing some wind-eroded clasts.

2.6 Unit B upper part, interpretation

The loose, heterogeneous character, variable clast fabric and folded sand laminae of the uppermost part of unit B suggest deposition by glacigenic sediment flow. The sorted laminae were deposited by sediment-laden meltwater sheet or rill flow on top of the sediment surfaces (Lawson 1989). Many of these laminae were incorporated and deformed during remobilization and flow of the water-saturated, already deposited diamictic sediments. The fact that sand laminae mimic the undulating B/C contact, which is shown below to be deformed by melting of buried ice, indicates a supraglacial origin for this part of unit B.

A moderate fabric is expected in the internal shearing zone of flows with a relatively high amount of water (Lawson 1982). This type of flow is also expected to erode channel banks and the eroded sediment can be incorporated as intraclasts (Lawson 1982).

The sand and gravel beds in the upper part of unit B are interpreted as glaciofluvial sediments deposited by running water in a stagnant-ice environment. Wind-eroded clasts, indicating subaerial conditions, were sometimes embedded in the glaciofluvial sediments. The angular chert fragments imply frost shattering of clasts in a

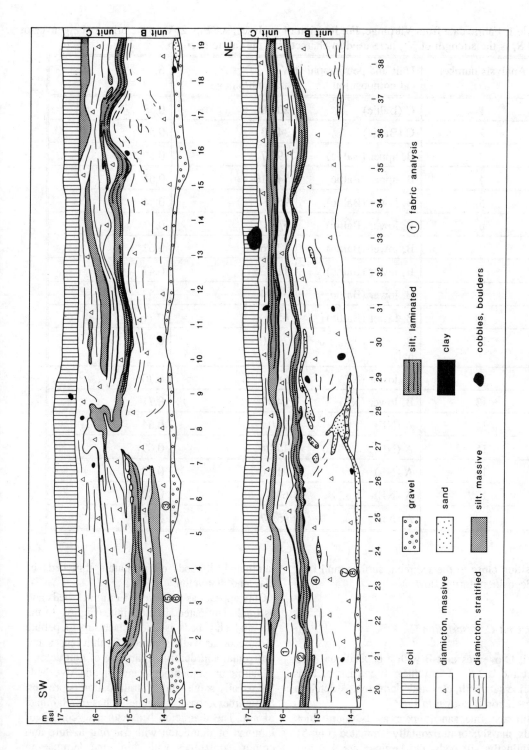

Fig. 3. Representative part of the Värpinge sections with units B and C.

Table 1. Fabric data from Värpinge. For location of analyses, see Fig. 2. V_1 is the principal eigenvector and S_1 is the strength of V_1, here used to characterize the fabric strength.

Analysis number	Unit and petrographical composition	V_1 azimuth/plunge	S_1
1	C (Baltic)	202/8	0.58
2	C (Baltic)	342/3	0.57
3	B, upper (Baltic)	192/7	0.69
4	B, upper (Baltic)	340/10	0.64
5	B, lower (Baltic)	95/5	0.54
6	B, lower (Baltic)	5/1	0.53
7	B, lower (Baltic)	120/6	0.62
8	B, lower (Baltic)	118/25	0.47
9	B, lower (Baltic)	351/4	0.57
10	B, lower (Baltic)	65/7	0.71
11	B, lower (transitional)	79/10	0.84
12	B, lower (NE)	58/10	0.70
13	B, lower (NE)	34/11	0.71
14	A (NE)	72/11	0.64
15	A (NE)	174/6	0.69
16	A (NE)	91/56	0.69
17	A (NE)	64/4	0.80
18	A (NE)	95/17	0.65

position close to the sediment surface, and later redeposition during flow.

2.7 Unit C, description

Unit C overlies unit B with a depositional contact and is 1.5 - 2.5 m thick. It consists of beds of fine sand, silt, clay and diamicton. A detail of the section is shown in Fig. 4.

Silt and fine sand appear as 1-20 cm thick beds, massive or horizontally laminated (Fig. 5). Within the silt beds, the laminae are 1-3 mm thick and normally graded. Sand beds are massive or either normally or inversely graded. The laminated silt beds often contain thin beds of clay and diamicton.

Clay appears as massive 1-5 cm thick beds and as finely laminated silt/clay units (Fig. 6). In the silt and clay beds isolated clasts up to pebble size occur, as well as intraclasts, thin lenses and horizontal and deformed laminae of diamicton.

Diamicton sometimes occurs in massive beds but usually some kind of stratification is present. Sometimes a massive diamicton has thin laminae of silt. The diamicton beds can also consist of laminae of diamicton with varying texture and colour, sometimes with thin sand laminae in between. The composition of the diamicton varies between clayey sandy diamicton and silty

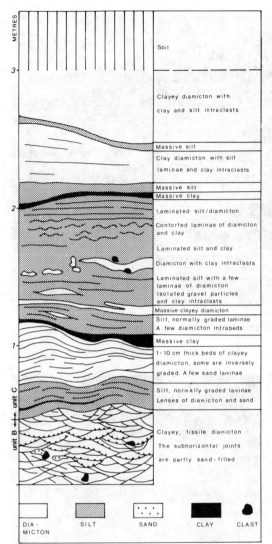

Fig. 4. The part between 2 - 3 m in Fig. 3 of the Värpinge section with detailed sediment descriptions.

clay diamicton. The diamicton beds often contain clay and silt intraclasts (Fig. 7).

Clast fabric was investigated at two places in stratified diamicton with laminae of silt and clay. Both fabric analyses have S_1 values lower than 0.60 and can be described as very weak to random.

Unit C has a Baltic petrographical composition, identical to the one in the upper part of unit B.

2.8 Unit C, interpretation

Unit C consists of lithofacies types which are typical of a proximal glaciolacustrine environment (e.g. Gustavson 1975a; Ashley 1989). The massive clay settled from suspension during periods when other deposition had ceased. The finely laminated silt and clay settled out of suspension from meltwater plumes transported into the basin as interflow or overflow currents (e.g. Ashley 1985).

The graded silt and sand was deposited by density underflows which produced a series of graded beds. The origin of these flows may have been high-density sediment-laden meltwater or turbidity currents generated by subaqueous slumping or debris flows.

The isolated pebbles are interpreted as icerafted particles. Intraclasts of diamicton were probably released from floating ice and embedded in the bottom sediment.

The diamicton beds in unit C are interpreted as deposited by cohesive debris flows in a subaqueous environment (Lowe 1982). The diamicton often consists of stacked beds with slightly variable composition, which represent successive flow units. The silt laminae between diamicton beds may have been deposited by settling from suspension between flow events or from turbidity currents generated by flow separation behind the debris flow (Hampton 1972).

The diamicton beds contain silt and clay intraclasts which could have been incorporated in the flow by erosion or be fragments of silt and clay beds from previously deposited sediment piles which were redistributed due to slumping or collapse of buried ice.

The random to weak fabric is typical of subaqueous debris flows (Kurz & Anderson 1979).

2.9 Deformation

Unit C is partly subhorizontal and partly deformed in an irregular sinous, undulating style (Figs. 2,3,8). The amount of deformation decreases upwards in unit C and the uppermost silt bed is almost horizontal. The deformation pattern of units B and C and the fact that individual beds can be followed throughout the section make it possible to reconstruct the deposition-deformation

Fig. 5. Laminated silt and fine sand of unit C with interbeds and intraclasts of diamicton. Frame height is 25 cm.

history of unit C in detail (Fig. 9).

3 RECONSTRUCTION OF THE DEPOSITIO-
NAL ENVIRONMENT

Our story begins with the deposition of the lodgement till of unit A. This was deposited at the base of an active ice, coming from the east. The ice then stagnated and supraglacial debris was redeposited as sediment flow diamictons ("flow till"). Sand beds were deposited by running meltwater between flow events. Melt-out till was formed below the stagnant ice. The horizontal position of the sand and gravel beds in the lower part of the melt-out till indicates that no post-sedimentary deformation took place. This suggests subglacial deposition (basal melt-out).

A change in sedimentary environment is marked by the deposition from density underflows and suspension that took place directly on top of the diamicton surface of unit B. This reveals a rising water level, which first affected the SW part of the section, which must have been the lowest-lying at that time. Unit C was draped over the supraglacial sediments, and deposition from debris flows, density underflows and suspension

Fig. 6. Heterogeneous, clayey diamicton of unit B draped by laminated silt and clay of unit C. A small slide in the lower part of the silt sequence has caused deformation of the overlying laminae. The silt bed is 20 cm thick.

Fig. 7. Unit C. Sandy diamicton is overlain by laminated silt and clay. The upper diamicton bed contains abundant clay intraclasts.

fall-out of clay and silt were the active processes.

The debris flows originated from the supraglacial sediments. Individual diamicton beds can be followed from unit B into unit C (Figs. 3,9). The sediments of unit C were gradually lowered when the buried ice melted and locally oversteepening of the basin floor triggered slumping and flow. At the same time, the englacial debris of the buried ice was gradually released to form melt-out till. On the whole, unit C was very gently lowered and the lowermost silt bed has not been disrupted except at 6-8 m where the concentration of englacial debris must have been very high (Fig. 9). The gradual melt-out of the buried ice is also illustrated by the decreasing amount of deformation upwards in unit C.

The presence of dropstones and diamicton lenses within the clay and silt beds of unit C indicates rafting from floating ice - icebergs, bergy bits or winter ice (cf. Gilbert 1990). The diamicton beds in the upper part of unit C require a source of poorly sorted debris. This could be stagnant, debris-covered ice or stagnant ice that rose from the basin-floor and started to float (cf. Gustavson 1975b). This floating ice could dump its debris on the bottom to be redeposited by debris flow.

Fig. 8. Unit B (lower dark diamicton) draped by silt and diamicton beds of unit C. Note the decreasing amplitude of deformation upwards. The section is 3.5 m high. The photo corresponds to 13 - 19 m in Fig. 3.

4 CONCLUSIONS AND DISCUSSION

The section at Värpinge comprises a sequence of sediments deposited under variable environmental conditions. Active ice, moving from the east, stagnated and during a subsequent transgression the disintegrating ice melted out completely.

This series of events has resulted in a lodgement till overlain by a melt-out till and supraglacially deposited flow diamicton ("flow till"). The glacial sequence is draped by glacioaquatic sediments deposited by sediment gravity flow and by fall-out from suspension. During deposition of the subaquatic sediments the buried, disintegrated ice gradually melted, which caused deformation of the overlying sediment pile. A reconstruction of the successive deposition/ melt-out/deformation phases was possible (Fig. 9) as the individual beds could be traced along the section.

The glacioaquatic unit C is stratigraphically a part of the Lund Diamicton because of its position on top of the youngest till in the area and the identical depositional environment of the sediment. It has been demonstrated that stagnant ice formed the eastern barrier of the basin where the Lund Diamicton was deposited (Malmberg Persson 1988; Malmberg Persson & Lagerlund 1990). The present investigation indicates that disintegrating ice was probably widespread in the area. On higher ground around Lund, however, the extremely well-preserved periglacial ground surface indicates ice-free conditions (Lagerlund 1987b; Malmberg Persson & Lagerlund 1990).

We conclude that before the transgression, the lower-lying areas in southwest Skåne still held stagnant, disintegrating ice and supraglacial and proglacial deposition took place. Higher ground, often with drumlinized morphology, was free of ice and subject to periglacial processes. This landscape was inundated up to at least 70-80 meters above sea level (Malmberg Persson 1988) and the Lund Diamicton was deposited as a draping glacioaquatic sediment on top of the partly deglaciated landscape.

Seen in a regional context, the series of events recorded at the Värpinge locality, with a last active ice movement from the east, a deglaciation sequence with a Baltic petrographical composition and a transgression of the area, can not be explained by the traditional till-stratigraphic model for Skåne (e.g. Holmström 1904; Ringberg 1989). For the area around Lund, this model predicts an uppermost basal till with Baltic clast content, deposited by ice coming from the southwest. This would be underlain by till and possible deglaciation sediments with NE petrographical composition and deposited by ice coming from northeast - east.

However, the evidence at Värpinge for a last active ice movement from the east is in agreement with other recent investigations from western and central Skåne (Adrielsson 1984; Åmark 1987; Lagerlund 1987a; Malmberg Persson 1988). The corresponding till unit has generally a NE-type petrographical composition (Åmark 1987; Malmberg Persson 1988), but in the westernmost parts of Skåne the same unit is characterized by rock types of Baltic provenance (Adrielsson 1984; Lagerlund 1987a). At the Värpinge section a gradual transition within the deglaciation sequence (unit B) from a NE to a Baltic petrographical composition was observed. The Baltic material is to a large extent derived from the southern part of the Baltic depression (Danian limestones) and does not originate from areas east of Värpinge, although ice flow was from the east.

Fig. 9. Cartoon showing processes and resulting deposits for successive stages in the environmental development at Värpinge.
1. Stagnant, disintegrating ice where supraglacial melting and release of debris results in deposition of diamictons and glacifluvial sediments. Basal melt-out till is released on top of already deposited lodgement till.
2. A regionally rising water level caused the iceblock to the left to become buoyant and drift away, leaving a depression that quickly fills with sediment gravity flow deposits.
3. The supraglacial sequence is being draped by sediments deposited from suspension, underflows and rafting.
4. Buried ice melts and the overlying sequence is being lowered and deformed. Rafting and deposition from suspension etc. are still active.
5. Water level drops and buried ice has melted. The deposits now correspond to the investigated section (Fig. 3).

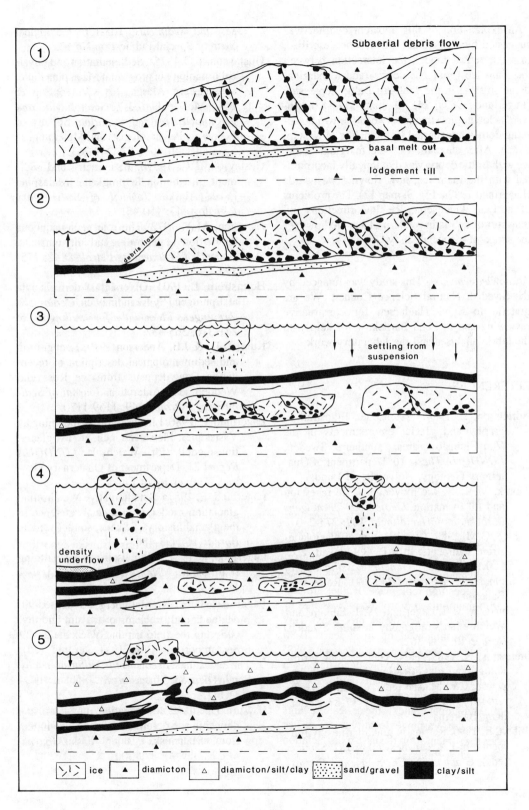

An explanation of this apparent conflict was suggested by Lagerlund (1987a) who described, in a theoretical model, the interaction between the main ice (with NE debris) and a marginal dome situated in the southern Baltic depression (Lagerlund 1987a: Fig. 9, map 11). A north-wards-flowing icestream between the two coalescing domes transported Baltic debris to western Skåne. After stagnation of the marginal dome, ice with Baltic debris was dynamically incorporated with the main ice, now flowing westwards (Lagerlund 1987a: Fig. 9, map 12). The problems of the Late Weichselian ice flow directions and transport of debris in western Skåne are however the scope of a special study.

Acknowledgements - This study was financed by the Swedish Natural Science Council. We are grateful to Sylvi Haldorsen for constructive reviewing and to Lena Barnekow for help with the fabric analyses and the laboratory work.

REFERENCES

Adrielsson, L. 1984. Weichselian lithostratigraphy and glacial environments in the Ven-Glumslöv area, southern Sweden. *LUNDQUA Thesis* 16. Department of Quaternary Geology, University of Lund.

Åmark, M. 1987. Ice movements, ice recession and till formation. *Geologiska föreningens i Stockholm förhandlingar* 109: 275-290.

Ashley, G.M. 1989. Classification of glaciolacustrine sediments. In R.P. Goldthwait & C. L. Matsch. (eds), *Genetic classification of glacigenic deposits*: 243-260. Rotterdam: Balkema.

Berglund, B. & E. Lagerlund 1981. Eemian and Weichselian stratigraphy in south Sweden. *Boreas* 10: 323-362.

Dreimanis, A. 1989. Tills: Their genetic terminology and classification. In R.P. Goldthwait & C.L. Matsch (eds), *Genetic classification of glacigenic deposits:* 17-83. Rotterdam: Balkema.

Gilbert, R. 1990. Rafting in glacimarine environments. In J.A. Dowdeswell & J.D. Scourse (eds), *Glacimarine Environments: Proces-*

ses and Sediments: 105-120. Geological Society Special Publication No 53.

Gustavson, T.C. 1975a. Sedimentation and physical limnology in proglacial Malaspina Lake, southeastern Alaska. In A.V. Jopling & B.C. McDonald (eds), *Glaciofluvial and glaciolacustrine sedimentation*. Society of Economic Paleontologists and Mineralogists Special Publication 23: 249-263.

Gustavson, T.C. 1975b. Bathymetry and sediment distribution in proglacial Malaspina Lake, Alaska. *Journal of Sedimentary Petrology* 45: 450-461.

Hampton, M.A. 1972. The role of subaqueous debris flow in generating turbidity currents. *Journal of Sedimentary Petrology* 42: 775-793.

Holmström, L. 1904. Öfversigt af den glaciala afslipningen i Sydskandinavien. *Geologiska föreningens i Stockholm förhandlingar* 26: 241-316, 365-432.

Kurz, D.D. & J.B. Anderson 1979. Recognition and sedimentological description of recent debris flow deposits from the Ross and Weddell seas, Antarctica. *Journal of Sedimentary Petrology* 49: 1159-1170.

Lagerlund, E. 1980. Litostratigrafisk indelning av Västskånes Pleistocen och en ny glaciationsmodell för Weichsel. *LUNDQUA Report* 21. Department of Quaternary Geology, University of Lund.

Lagerlund, E. 1987a. An alternative Weichselian glaciation model, with special reference to the glacial history of Skåne, south Sweden. *Boreas* 16: 433-450.

Lagerlund, E. 1987b. Weichselisens avsmältning från Skåne. *Svensk Geografisk Årsbok* 1987: 9-26.

Lagerlund, E. & K. Malmberg Persson 1990. Methods and problems of till stratigraphy. Guide to the field trip in SW Skåne 28-29 September 1988. In E. Lagerlund (ed), *Methods and Problems of Till-Stratigraphy - INQUA-88 proceedings*: 50-58. LUNDQUA report 32.

Lawson, D.E. 1982. Mobilization, movement and deposition of active subaerial sediment flows, Matanuska Glacier, Alaska. *Journal of Geology* 90: 279-300.

Lawson, D.E. 1989. Glacigenic resedimentation: Classification concepts and application to mass-movement processes and deposits. In R.P. Goldthwait & C.L. Matsch (eds), *Genetic classification of glacigenic deposits*: 147-169. Rotterdam: Balkema.

Lowe, D.R. 1982. Sediment gravity flows: II. Depositional models with special reference to the deposits of high-density currents. *Journal of Sedimentary Petrology* 52: 279-297.

Malmberg Persson, K. 1988. Lithostratigraphic and sedimentological investigations around the eastern boundary of Baltic deposits in central Scania. *LUNDQUA Thesis 23*. Department of Quaternary Geology, University of Lund.

Malmberg Persson, K. & E. Lagerlund 1990. Sedimentology and depostitional environments of the Lund Diamicton, southern Sweden. *Boreas* 19: 181-199.

Mark, D.M. 1973: Analysis of axial data, including till fabrics. *Geological Society of America Bulletin* 84: 1369-1374.

Smith, N.D. & G.M. Ashley 1985. Proglacial lacustrine environment. In G.M. Ashley, J. Shaw & N.D. Smith (eds), *Glacial sedimentary environments*: 135-216. SEPM short course No. 16.

Ringberg, B. 1989: Upper Late Weichselian lithostratigraphy in western Skåne, southernmost Sweden. *Geologiska föreningens i Stockholm förhandlingar* 111: 319-337.

Formation and Deformation of Glacial Deposits, Warren & Croot (eds) © 1994 Balkema, Rotterdam, ISBN 90 5410 096 6

The deglaciation history of the Hofsárdalur valley, Northeast Iceland

Thorsteinn Sæmundsson
Department of Quaternary Geology, Lund University, Sweden

ABSTRACT: Contradictory interpretations of the timing and pattern of the last deglaciation of Iceland have highlighted the need for data on the deglaciation history of northeastern Iceland, which have so far been poorly known. The Hofsárdalur valley, in the Vopnafjörður area, was investigated with regard to glacial morphology, lithostratigraphy and chronology to reconstruct the glacial development. It is concluded that glaciers extended beyond the present coast until the last stage of the Late Weichselian. Possibly the glacial retreat was first interrupted near the present outer coast, simultaneously with a relative sea-level of 55 m above the present. Later the ice margin retreated to a position approximately 11-12 km inside this coast, where a 3 km long zone of stratified drift deposits mark a stillstand or small advance. The relative sea level during this episode is marked by a set of strandlines rising from approximately 45 m in the east to 70 m close to the ice marginal position. This glacial event has been dated to approximately 9,900 - 9,600 BP. The deglaciation history of Hofsárdalur suggests extensive glaciation of the coastal areas and high relative sea levels well into the Preboreal.

1 INTRODUCTION

The Hofsárdalur valley is situated in the Vopnafjörður district, between Bakkaflói in the north and Héraðsflói in the south, in NE-Iceland (Fig. 1). It is the southeasternmost and longest of three SW-NE trending valleys in the area. Hofsárdalur has two tributary valleys, Sunnudalur and Fossdalur. On the southern side of the valley the Smjörfjöll mountain massif rises steeply to 900-1000 m but on the northern side there is only a low ridge, Hraunlína gradually rising from about 180 m a.s.l. near the valley mouth to about 300 m a.s.l. near the mountain Burstarfell (Figs. 1 & 2). The Hofsá river drains Hofsárdalur and has its source in the highlands to the south. On its way through Hofsárdalur it is joined by several tributaries, mainly from the southern side. Hofsárdalur is broad and shallow, about 24 km long and opens into the Vopnafjörður fjord. The valley floor, which rises from sea-level to 60-70 m a.s.l. at the head, is to a large extent covered by glacial and marine sediments of Late Weichselian or early Holocene age. These sedi-

ments form extensive terraces along both sides of the valley (Fig. 2), whereas the valley floor is mostly covered by recent fluvial deposits.

The bedrock in the area belongs mainly to the Tertiary (Miocene-Pliocene) Basalt Formation, but in the west younger (≥ 0.7 m.y.) lava formations can be found (K. Sæmundsson 1974, Jóhannesson & K. Sæmundsson 1989). The Tertiary basalts are mainly subaerial tholeiitic lava flows with red or red-brown clastic interbeds. Some intermediate and acidic rocks occur in the mountains to the south. The Tertiary lava flows are heavily intersected by dykes, with a general N-S trend.

The present paper describes the deglaciation history of Hofsárdalur. It is the first publication of results from the Vopnafjörður project, which aims to get a clear and well dated picture of the deglaciation pattern in the district - including the history of shore-line displacement and valley fill sedimentation. Hofsárdalur was selected as a study area for the deglaciation due to its relatively continuous and dateable sedimentary sequence.

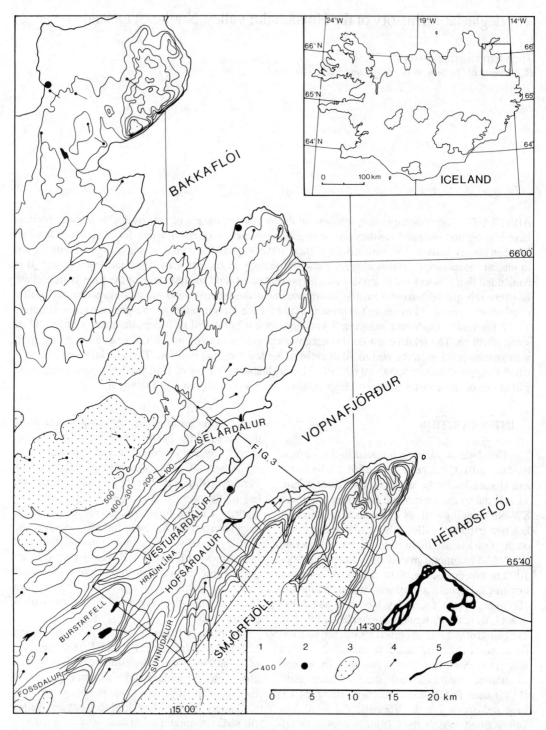

Fig. 1. Location map of the Vopnafjörður district. (1) Contour lines, 100 m interval, (2) towns, (3) areas above 500 m, (4) glacial striae, direction towards the dot, (5) streams and lakes.

2 THE ICELANDIC DEGLACIATION CONCEPTS

Two concepts have been presented for the last deglaciation of Iceland. An earlier one was published by Einarsson (1961, 1964, 1967, 1968, 1973, 1978, 1979, Einarsson & Albertsson 1988). According to that concept almost the whole island was covered by glaciers at the maximum stage of the last glaciation (see also Ashwell 1975, Andersen 1981), and the ice-sheet reached far beyond the present coasts. According to Ólafsdóttir (1975) moraines, probably marking the maximum glaciation, are found on the shelf off central western Iceland, at 200-250 m depth. Einarsson (1973, 1978, 1979) defined two interstadials during the general ice retreat, Kópasker and Saurbær, which he correlated with the Bölling and Alleröd interstadials in NW-Europe (Mangerud et. al. 1974), and two stadials, Álftanes and Búði which he correlated with the Older and Younger Dryas stadials in NW-Europe (Fig. 3). Einarsson's concept has been questioned and revised during the past few years, on the basis of new data from different parts of Iceland; e.g. Ingólfsson (1985, 1987, 1988), Pétursson (1986), Hjartarson & Ingólfsson (1988), Hjartarson (1989) and Norðdahl & Hafliðason (1990). These new data have changed the concept of ice extent during the Younger Dryas and Preboreal times by showing that the ice-extent during especially the Younger Dryas, but also during the Preboreal, was considerably larger than earlier anticipated. According to this new concept the age of the most recent marine maxima has to be re-evaluated. Hjartarson & Ingólfsson (1988) have showed that the Búði stage moraines in southern Iceland (Fig. 3F) are of Preboreal but not Younger Dryas age. This has highlighted the need for new data on the deglaciation history of northeastern Iceland.

The late-glacial history of northeastern Iceland can still be considered poorly known except for Melrakkaslétta in the north. Pétursson (1986) investigated the type locality for the Kópasker interstadial (Fig. 3C). There, glaciomarine sediments of Kópasker age (^{14}C dated to 12,600 BP) are overlain by till. Pétursson (1986) stated that this till is of Younger Dryas age, and not of Older Dryas age as Einarsson suggested earlier. Pétursson's conclusion was based on the fact that the till contained shell fragments of early Alleröd age. Pétursson dated the final glacial retreat from the area to about 10,100 BP.

An extensive reconnaissance study of eastern and southeastern Iceland (Fig. 3E) was made by Norðdahl & Einarsson (1988) to reconstruct the history of deglaciation and sea-level changes. They recognized three separate stages in the ice retreat, the Fáskrúðsfjörður stage, the Breiðdalur stage and the Berufjörður stage, but were not able to establish the absolute age of these stages and their related marine levels. However, they made a tentative comparison with southern and southwestern Iceland and suggested that the formation of the Fáskrúðsfjörður moraines occurred concurrently with the Álftanes stadial, the Breiðdalur moraines with the Búði stadial and that the Berufjörður moraines might be of Early Holocene age. But they also noted the possibility that the Berufjörður moraines could be slightly older, and then reflect a twofold Younger Dryas stadial.

3 PREVIOUS INVESTIGATIONS OF THE GLACIAL HISTORY OF VOPNAFJÖRÐUR

The first geologist to study the Quaternary in the Vopnafjörður area was Thoroddsen (1904, 1905-06, 1911). He studied changes of relative sea-level in the area and described conspicuous terraces of sand and gravel lying on top of fine-grained sediments, which in the outermost Selárdalur and Vesturárdalur valleys reach 50 m a.s.l. In Hofsárdalur he described large terraces running along both sides of the valley, mainly consisting of fluvioglacial sediments. He also mentioned large block fields in Hofsárdalur and claimed that the valley had once been a fjord.

Jux (1960) also described terraces in Vopnafjörður at 40-60 m a.s.l., which he assumed to have been formed during the Alleröd chronozone. In one section at Fell, on the northern side of Hofsárdalur, at 40-50 m a.s.l. he found a 2-3 m thick layer of coarse gravel on top of a 15-20 m thick grayish silty sediment, containing marine shells.

Einarsson (1961, 1968, 1973) briefly mentioned the terraces in the Vopnafjörður area, at about 40 m a.s.l., and claimed that they correlated with the highest marine limit in Iceland.

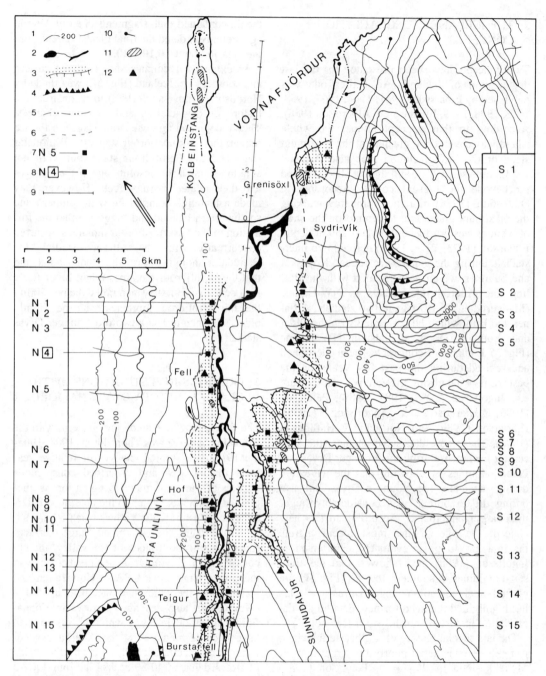

Fig. 2. Glacial landforms and location map of the Hofsárdalur valley. (1) Contour lines, 100 m interval, (2) lakes, streams and rivers, (3) terraces, (4) escarpments, (5) older marine limit, (6) younger marine limit, (7) location and number of sections; N=northern side of valley, S=southern side of valley, (8) location and number of ^{14}C-dated section, (9) location of transect in Figs. 5 and 6, scale in km, (10) glacial striae, direction towards the dot, (11) outcrops of bedrock, (12) farms.

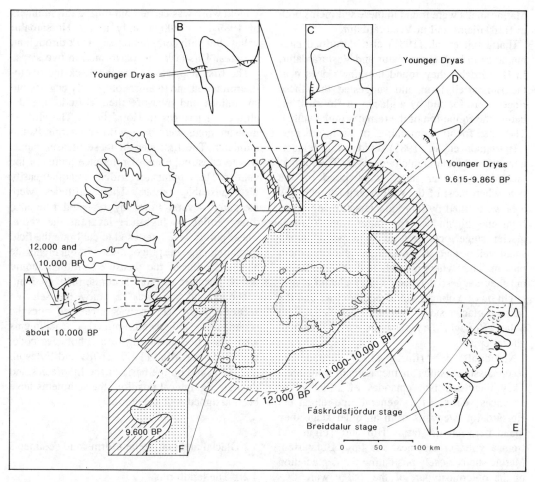

Fig. 3. Einarsson's (1978) reconstruction of the 12,000 BP (Álftanes stadial) and 11,000-10,000 BP (Búði stadial) ice front isochrones and the new concept of the Late-glacial ice-extent from different parts of Iceland, from e.g. A: Ingólfsson (1985, 1987, 1988), B: Norðdahl (1983) and Norðdahl & Haflidason (1990), C: Pétursson (1986), D: This work, E: Norðdahl & Einarsson (1988), F: Hjartarson & Ingólfsson (1988).

K. Sæmundsson (1977) outlined the main geo-morphological features of the Vopnafjörður area on his geological map of NE-Iceland. On the Kolbeinstangi peninsula (Fig. 2) he marked the marine limit at 50 m a.s.l.

Hjartarson et. al. (1981) also studied the Vopnafjörður area. They estimated the surface altitude of some larger terraces. In Selárdalur they found two generations of terraces, at about 50 and 55 m a.s.l. in the outer part of the valley, and claimed that their altitudes rose upvalley to 60 and 65 m a.s.l., respectively. They noted that the terraces in Selárdalur contain much less silt and are smaller than those in Vesturárdalur and Hofsárdalur. In Vesturárdalur, they traced a terrace upvalley where its altitude reached about 90 m a.s.l. In Hofsárdalur they measured the altitude of terraces which rise from 40 m a.s.l. in the coastal areas on the southern side of Vopnafjörður to 80 m a.s.l. near the farm Teigur (Fig. 2). The terraces in Hofsárdalur are mainly of silty material, usually with gravel on the top.

Marine shells were found in these sediments both in Hofsárdalur and in Vesturárdalur.

Hjartarson et. al. (1981) also described end-moraines in both Hofsárdalur and Vesturárdalur. In Hofsárdalur they found moraine ridges near the Burstarfell farm and suggested that these ridges were formed by a glacier in Hofsárdalur, before the formation of the terraces in the valley. They also found endmoraines in Vesturárdalur.

Hjartarson et. al. (1981) divided the deglaciation in East Iceland into two stages. The former, the "Valley-glacier stage", corresponds to the time when most of the valleys and fjords in the area were filled by large glaciers flowing down from the highlands. The latter, the "Cirque-glacier stage", corresponds to a period when small valley glaciers and cirque glaciers where present. They where not able to date these stages but they suggest that the "Valley-glacier stage" might be of Younger Dryas age or older and the "Cirque-glacier stage" might be correlated with a late Younger Dryas or an early Preboreal glacial episode.

Norðdahl & Hjort (1987) made a reconnaissance study in the Vopnafjörður area. They identified two episodes of stillstand or advances during the general ice-retreat in Hofsárdalur. Deposits from an older stage they found between the farms Hof and Teigur and from a younger one near the farm Burstarfell. Marine shells closely postdating the deglaciation of the outermost part of the valley were ^{14}C-dated to 9,980 BP and 10,230 BP (dates not corrected for reservoir age). Norðdahl and Hjort (1987) suggested that the most advanced position of the ice inside the present coast and the maximum sea-level in the area dates from the Younger Dryas chronozone and later stillstands or advances probably occurred during the Preboreal.

4 THE PRESENT STUDY

4.1 Methods

The study in Vopnafjörður started with air photo interpretation, including classification of the main landforms, elevated shorelines and selection of sections for sedimentological and stratigraphical studies.

Field work was carried out during the summers of 1989 and 1990, mainly in the Hofsárdalur valley, but with complementary work throughout the area. This task was performed in two steps:

The first step was to groundcheck the photo-interpretation and to morphologically classify old strandlines and measure their altitude and the altitude of terraces in Hofsárdalur. The altitude measurements were done with an anergic Paulin altimeter. The first step in these measurements was to construct a net of reference points in the valleys. These points are bridges and other easily distinguishable points. Their altitudes were measured several times by repeated reference after zeroing to high-tide level at the shore. These points were later used to calibrate the field measurements. In spite of these arrangements the absolute value of the measured altitude should not be regarded as more accurate than +/- 3 m.

The second step was to study the selected sections. Each section was measured and the sediments described and sampled. The sediments were classified according to a lithofacies code, modified after Miall (1977, 1978) and Eyles et. al. (1983) and on the basis of the lithofacies and their stratigraphical position, the sediments have been assigned to 7 categories.

4.2 Glacial and marine landforms and sediments

4.2.1 The landforms.

The main geomorphological features of the glacial and marine landforms in Hofsárdalur are outlined in Fig. 2. The most distinctive landforms are extensive terraces at variable altitudes, stretching from the coastal areas towards the head of the valley (Fig. 2). Partly they are raised beaches and raised abrasion terraces, marking ancient sea-levels in the outer part of the valley, but mostly they are remnants of valley fill sediments deposited during a general lowering of relative sea-level after deglaciation. Later they were fluvially eroded.

Two sets of ancient shorelines, and of ice-front positions connected with them, are found in the central and outer parts of Hofsárdalur. The older one, which defines the maximum marine limit at about 55 m a.s.l., is only present in the coastal area northeast of the Grenisöxl farm and on the

Kolbeinstangi peninsula, just north of the mouth of Hofsárdalur (Fig. 2).

A younger marine limit is found in the outer and central parts of the valley, from the area inside Grenisöxl upvalley towards the area between Hof and Teigur (Fig. 2). On the southern side of the valley this marine limit can be followed continuously on the terrace surfaces, some 11-12 km upvalley, from about 45 m a.s.l. to 65-70 m a.s.l. On the northern side of the valley the younger marine limit can be found as discontinuous raised beaches (Fig. 2).

The terraces in the outer parts of the southern side of the valley, reaching some 6 km inland, occur at one elevation (Fig. 2). These terraces are mainly composed of stratified sandy gravel, forming small deltas along the side of the valley. Further upvalley the deltas become larger, and are mainly composed of stratified sand. Their surfaces show marks of several younger beach platforms and abrasion terraces.

The terraces in the area between the farms Hof and Teigur and at the mouth of Sunnudalur (Fig. 2), rise from 70 m a.s.l. near Hof to 80 m a.s.l. near Teigur. At Sunnudalur the surface consists of boulder-rich coarse-grained glaciofluvial sediments with fluvial channels and kettle-holes.

The area west of Teigur in Hofsárdalur as well as in Sunnudalur are covered by glaciofluvial deposits, forming terraces on both sides of the valleys. On the northern side of Hofsárdalur these sediments form terraces at several elevations but on the southern side only one elevation can be found.

The northern valley slope in the inner part of Hofsárdalur, rises steeply from the highest terraces at about 90-100 m a.s.l. up to about 500 m a.s.l. The lower part of the slope is covered by talus. Further downvalley, on the slopes between the Burstarfell and Teigur farms, kame terraces and lateral channels occur at different heights. The southern valley slope is not as steep as the northern one, and it is mantled by a relatively thin till cover.

4.2.2 The sediments.

The sediments in Hofsárdalur have been studied in numerous sections on both sides of the valley (Fig. 2). The five sections in Fig. 4 outline the main stratigraphical features of the sedimentary sequence in Hofsárdalur. These sections were selected as key-sections for their representation of sedimentation during different stages of the deglaciation. Their stratigraphical position is shown in Fig. 5.

Section S-1 (Fig. 4) is situated on the southern side of the valley northeast of the Grenisöxl farm (Fig. 2). It is about 7 m thick, mainly composed of parallel stratified sand (Ss) and sandy gravel (Gs), dipping 10° towards north and northeast.

Section N-4 (Fig. 4) is situated on the northern side of the valley, near the farm Fell (Fig. 2). The lowest unit in this section is massive diamicton (Dmm) with abundant angular to subrounded clasts. The diamicton is unconformably overlain by 15 m laminated silt and fine sand (Fl), with occasional 10-15 cm thick interbeds of coarser sand (Ss). Marine shells occurs in the sorted sediments. Discordantly on top of the fine sediments a 6 m thick stratified sand bed (Ss, St) is found.

Section S-4 (Fig. 4) is situated on the southern side of the valley (Fig. 2). It is 15 m high and composed of stratified coarse-grained sandy gravel (Gs) and coarse sand (Ss). The uppermost 3 m of this section consist of parallel stratified topset beds which dip 7-10° towards north and northeast and distally grade into foresets dipping 25-35° towards north.

Sections N-10 and N-11 (Fig. 4) are situated on the northern side of the valley in the area between the farms Hof and Teigur (Fig. 2). Section N-10 is about 40 m thick. It is composed of a 37 m thick coarse-grained stratified dift deposit, overlain by 3 m of cross-stratified sand (St) and sandy gravel (Gt). The lowermost 30 m of this section are covered but in several small excavations the same coarse-grained sediments were observed. Section N-11 is about 30 m thick. The lowermost 20 m of this section consist of laminated silt and fine sand (Fl). Discordantly, the fine sediments are overlain by 10 m of cross-stratified sand (St) and sandy gravel (Gt).

The sections outlined in Fig. 5 show the stratigraphy of Hofsárdalur. Location of sections and the transect line are shown in Fig. 2. The sediments were divided into 7 categories, on the basis of lithofacies type and stratigraphic position. The following description of the units illustrates the lithofacies type, the thickness and

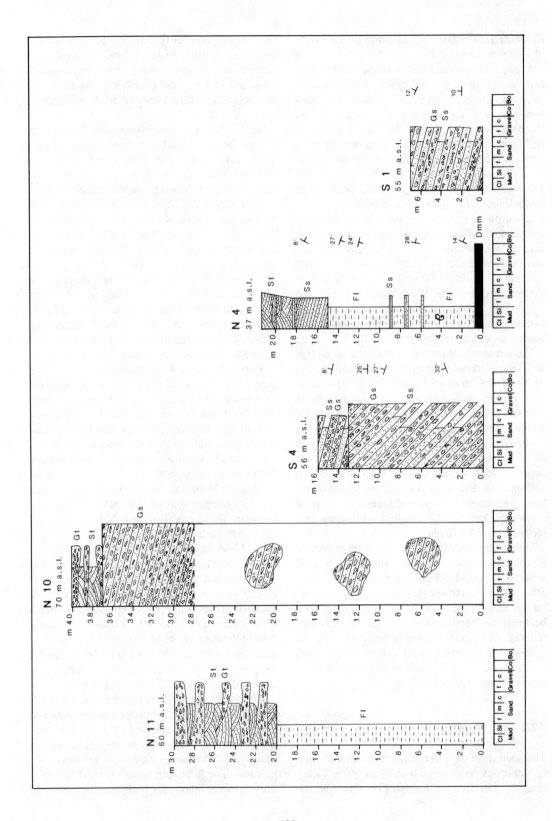

the spatial distribution of the sediments.

Unit A = till. This is the lowermost stratigraphic unit, resting on striated bedrock. It is a compact, massive diamicton (Dmm), with variable amounts of angular to subrounded pebbles, cobbles and boulders, in a sandy and silty matrix. Several striated cobbles and boulders occur within this unit. The average exposed thickness is 40-60 cm, the maximum thickness about 1 m.

Unit B = the older delta sediments. This unit occurs in a limited area northeast of Grenisöxl (Figs. 2 & 5, section S-1) where its surface reaches 55 m a.s.l. It is composed of stratified sand (Ss) and sandy gravel (Gs) with a maximum clast size of 20-30 cm. The sediments are parallel stratified and gently sloping (10°) towards north and northeast. This unit is poorly exposed, except in the about 7 m deep Grenisöxl gravel pit.

Unit C = glaciomarine silt. This unit is widespread in the outer parts of the valley, mainly on the northern side, and overlying unit A there (Fig. 5). It consists of a compact laminated sandy silt (Fl) with occasional interbeds of coarser sand (Ss). Subfossil mollusc shells and casts of mollusc shells occur in this unit near the farm Fell (Figs. 4 & 5 section N-4). The thickness of this unit varies, but the maximum thickness is about 15 m in the upvalley part and it gets thinner downvalley (Fig. 4).

Unit D = the younger delta sediments. This unit is found on the southern side of the valley, from the farm Syðri-Vík and some 8 km upvalley where it overlies units A and C (Fig. 5, sections S-2 to S-8). Its composition ranges from stratified coarse gravel (Gs) to stratified coarse and fine sandy sediments (Ss, St, Sr) mostly deposited as small deltas along the valley side. The thickness of this unit varies from a few meters to 15-20 m (Fig. 4).

Unit E = stratified drift (sub-unit Ea) interfingered with fine-grained sediments (sub-unit Eb). This unit is found in the area between Hof and Teigur in Hofsárdalur (Fig. 2) and also at the mouth of Sunnudalur. It is composed of coarse-grained stratified drift sediments, sub-unit Ea, (sections N-9, N-10, N-12, N-14, S-12 and S-14, in Fig. 5) and fine grained sediments, sub-unit Eb (sections N-11, N-13 and S-13, in Fig. 5). This unit was formed by meltwater from glaciers in Hofsárdalur and Sunnudalur (Fig. 3). Along the northern side of Hofsárdalur the coarser part of this unit (sub-unit Ea) forms three ridges, partly buried in finer-grained sediments (sub-unit Eb) (Fig. 6). Similar stratigraphic associations also occur on the southern side of Hofsárdalur (Fig. 5) and at the mouth of Sunnudalur.

The proximal sides of the two outermost course grained ridges, sections N-10 and N-12 (Figs. 5 & 6) consist of boulder-rich sandy gravel (Gs), but distal sides consist of finer sandy gravel and sand sediments. The proximal deposits are overlain by laminated silt and fine sand sediments (sections N-11 and N-13) which reach 30 m in thickness (Fig. 5). The fine grained deposits, in section N-13 (Fig. 5), consist of sand (Sh, Sr, St) interstratified with silt and fine sand (Fl). The innermost coarse-grained ridge, section N-14, consists of stratified boulder-rich gravel (Gs) and sand (Ss, St), with signs of deformation associated with kettle hole formation. The thickness of this unit is about 30-40 m (Fig. 6).

Unit F = the outer sandur sediments. This unit is widespread on the northern side of the valley from the area between Hof and Teigur and some 11-12 km downvalley (Fig. 5, sections N-13 to N-1 upper part). It also occurs at the mouth of Sunnudalur. The unit is very complex in both texture and structure and its thickness gradually decreases upvalley (Fig. 5). The lower part of this unit consists of sandy sediments (Ss, St, Sr) with occasional pebbles and cobbles and lenses of fine gravel, where its upper part mainly consists of more or less sand (Ss, St) and sandy gravel (Gs, Gt).

Unit G = the inner sandur sediments. This unit

Fig. 4. Five key-sections selected from the sedimentary sequence in Hofsárdalur and representing different stages during the deglaciation of the valley. Locations of the sections are shown in Fig. 2 and their stratigraphical position in Fig. 5. The facies codes are modified from Miall (1977, 1978) and Eyles et. al. (1983). (Fl: Fines (silt, clay), laminated; Ss: Sand, stratified, (planar-parallel bedding); St: Sand, stratified, (trough cross-stratification); Gs: Gravel, stratified, (clear stratification); Gt: Gravel, stratified, (trough crossbeds)).

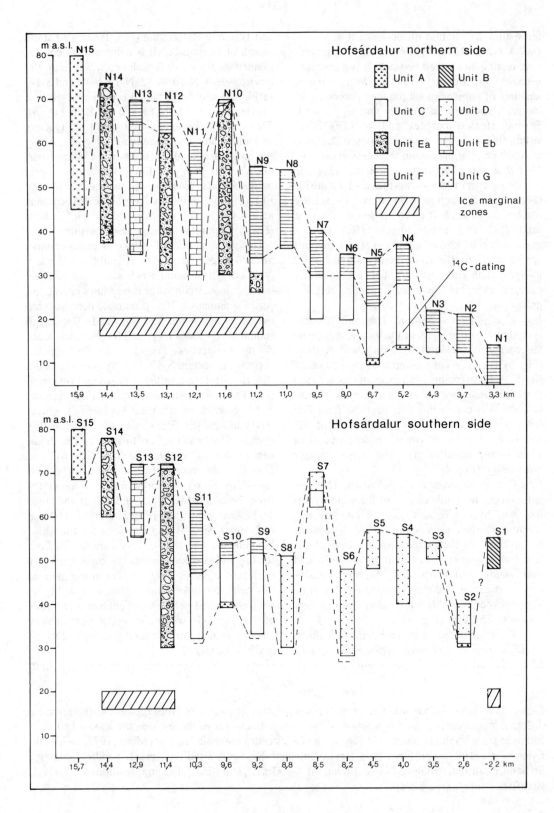

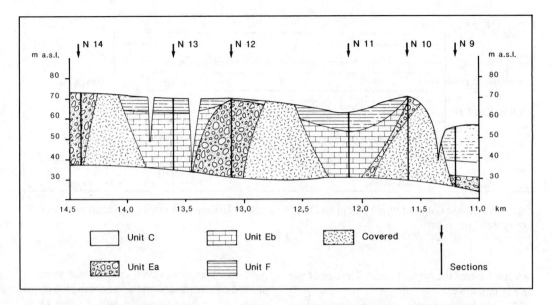

Fig. 6. Stratigraphical features of the 3 km wide zone of stratified drift deposits (Unit E), sections N-8 to N-14 (Fig. 5), in the area between Hof and Teigur in central Hofsárdalur. The location of sections is shown in Fig. 2.

is found in innermost Hofsárdalur (sections N-15 and S-15, Fig. 5) and in Sunnudalur. Its structure and composition are highly variable, both laterally and vertically and the grain size ranges from fine sand to boulder-rich gravel. Generally the coarseness increases upvalley.

5 INTERPRETATION

The stratigraphy of Hofsárdalur indicates a relatively continuous sedimentation, following the ice margin upvalley during its retreat from the coastal areas towards the head of the valley (Fig. 5).

The lower contact of the sedimentary sequence in Hofsárdalur is striated bedrock, which shows an iceflow direction parallel to the valley (Fig. 2). The lowest unit in the sedimentary sequence, resting on striated bedrock, is a till (Unit A). It was presumably deposited when the glaciers

Table 1. Radiocarbon dates of shells from the Hofsárdalur valley (from Norõdahl & Hjort 1987).

Location (Fig 2)	^{14}C-date, years BP	Ref.no. of date	Altitude m a.s.l.
N-4 Fell	9865+/-90	Lu-2674	18 m
N-4 Fell	9615+/-70	Lu-2673	21 m

The samples were referred to 0,95 NBS oxalic acid standard, using the value of 5568 for the half life of ^{14}C. The base year is 1950. Corrections for ^{13}C/^{12}C ratios have been made, as have corrections for a reservoir age of 365+/-20 years (Håkansson 1983).

Fig. 5. Vertical sections from the northern and southern sides of Hofsárdalur valley. The position of the sections and the transect line are indicated on Fig. 2. Note that distances between sections are not to scale.

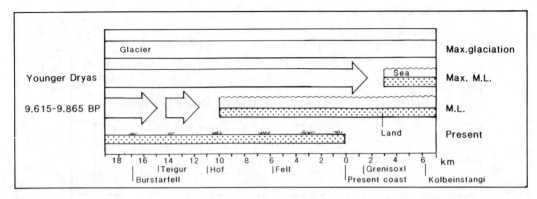

Fig. 7. Schematic diagram of ice extent and marine limits in Hofsárdalur during different stages in the Late-glacial ice retreat.

reached beyond the present coast. The age of this unit is unknown.

The first indication of a standstill or minor re-advance during the glacial retreat in Hofsárdalur has been found in the coastal areas on the southern side of the valley, northeast of Grenisöxl (Fig. 2). There delta deposits (Unit B) mark the maximum marine limit in the area at 55 m a.s.l. This delta seems to have little or no direct relation with other sediments in the valley. Most likely it was deposited near the ice margin, probably by a lateral stream. No other glacial sediments belonging to this stage have been found in the valley, but shore-lines at the older marine limit on the Kolbeinstangi peninsula, at 50 m a.s.l. (Fig. 2), show that this peninsula was ice-free at that time. Other evidence for this ice marginal position is that only 1-2 km inside Grenisöxl (Fig. 2) the marine limit lies at least 10 m lower. The maximum marine limit in the area has not yet been dated, but on basis of the similar degree of weathering as of the younger sediments and the necessarily rather high sea-level (late glacial transgression?) I tentatively suggest it to be of Younger Dryas age.

Later the ice margin retreated to the area between Hof and Teigur and at the mouth of Sunnudalur, approximately 11-12 km inside the present coast (Fig. 2). There, a zone, about 3 km wide with coarse-grained stratified drift (Unit E), indicates a stillstand in the ice retreat and possibly a small re-advance (Fig. 5, sections N-8 to N-14). Concurrently with the retreating ice margin the sea inundated the valley floor. The position of the sea-level during this stage is marked by strandlines on the southern side of the valley, rising from approximately 45 m a.s.l. in the coastal areas to 65-70 m close to this ice marginal position (Fig. 2). Thick, laminated silts and fine sands (Unit C) were deposited during this phase, mainly on the northern side where they overlie unit A (Fig. 5). Shells of Hiatella arctica and Balanus have been found in this formation near the farm Fell (Fig. 2). Radiocarbon dates place the deposition between about 9,865-9,615 BP, using the dates of Norðdahl & Hjort (1987) corrected for reservoir age (Table 1). Concurrently with the deposition of the fine (Unit C) sediments on the northern side, coarser sediments (Unit D) were deposited on the southern side, overlying Unit C there. These glacio-fluvial sediments, which to a large extent originate in the mountains to the south, form small deltas along the southern side of the valley. They occur in the area around the Syðri-Vík farm and some 9 km upvalley (Fig. 5). The delta topset beds indicate the younger marine limit, rising from 45 m a.s.l. to 65-70 m a.s.l. (Fig. 2). After this stage of stillstand and or minor re-advance the ice retreat continued, simultaneously with a lowering of the relative sea-level. Large quantities of sands and gravels were deposited on top of some older sediments in the outer parts of the valley (Unit F) and in the inner parts of the valley (Unit G) while relative sea-level was still relative high. Minor indications of stillstands or small advances have been found also in these sediments in upper Hofsárdalur. As relative sea-level fell towards present sea-level, the glacial rivers began to erode the older parts of the sedimentary sequence, leaving terraces at different heights along both sides of the valley. Deposition of coarser glaciofluvial sediments on top of the finer ones also took place during this stage (younger parts of Unit F).

6 CONCLUSIONS

In summary the history of deglaciation and sea-level change in Hofsárdalur is as follows (Fig 7):

1. During the maximum glaciation in Iceland the whole Vopnafjörður area was covered by glaciers which reached beyond the present coast. It has been suggested that the mountain tops south of Vopnafjörður were ice free during this stage (Thórarinsson 1937, Einarsson 1961, Steindórsson 1962, Sigbjarnarson 1983).

2. The first indication of a stillstand during the deglaciation has been found on the southern side of the valley in the outermost parts of Hofsárdalur. There a delta indicates an ice front position when sea-level stood at the maximum marine-limit in the area, at 55 m. No dating of this stage has been possible, but the ^{14}C-dates from the younger sediments in the valley indicate that it took place before about 10,000 BP, and it is tentatively suggested to be a Younger Dryas episode. In connection with the ice margin retreat from this position, the relative sea-level dropped at least 10 m.

3. Then the ice margin retreated to another position approximately 11-12 km inside the present coast, in the area between Hof and Teigur, where a 3 km wide zone of stratified ice-contact deposits marks another stillstand or a small re-advance. Relative sea-level during this stage is marked by strandlines rising from 45 m a.s.l. near the present coast to 65-70 m near the former ice margin. This stage has been radiocarbon dated to between about 9,865 - 9,615 BP approximately (Norðdahl & Hjort 1987).

4. The continuing ice retreat was uninterrupted, during a period of a general lowering of sea-level. Extensive sandur deposit accumulated in the inner parts of the valley, and also on top of the older deposits downvalley.

The deglaciation history of Hofsárdalur outlined above suggests extensive glaciation of the coastal areas in Late Weichselian times. A still-stand or a re-advance of the glacier in Hofsárdalur and the maximum marine limit in the valley tentatively dates from the Younger Dryas chronozone. A later stillstand and/or a small advance occurred during the Preboreal.

When comparing the data from Hofsárdalur with the deglaciation scenarios from the rest of Iceland (Fig. 3) it is evident that the ice extent in the area during the Younger Dryas chronozone was much greater than Einarsson (1973, 1978, 1979) suggested. The data fit much better with newer information from SW-, W- and N-Iceland, which indicates considerably heavier glaciation in Younger Dryas and Preboreal times (Ingólfsson 1985, 1987, 1988, Pétursson 1986, Hjartarson & Ingólfsson 1988, Norðdahl & Hafliðason 1990, Hjartarson 1989).

ACKNOWLEDGEMENTS

I would especially like to thank my supervisors Docent Christian Hjort, at the University of Lund and Dr. Hreggviður Norðdahl, at the University of Iceland, for proposing this investigation and for valuable and improving comments on the manuscript.

REFERENCES

Andersen, B.G. 1981: Late Weichselian Ice Sheets in Eurasia and Greenland. *In* G.H. Denton & T.J. Huges (eds.), *The Last Great Ice Sheets,* 1-65. John Wiley and Sons Inc. New York.

Ashwell, I.Y. 1975: Glacial and late glacial processes in Western Iceland. *Geografiska Annaler 57,* 225-245.

Einarsson, Th. 1961: Pollenanalytishe Untersuchungen zur spät- und postglazialen Klimageschichte Islands. *Sonderveröffentlichungen des Geologischen Institutes der Universität Köln 6.* 52 pp.

Einarsson, Th. 1964: Aldursákvarðanir á fornskeljum (English summary: Radiocarbon dating of subfossil shells). *In* G. Kjartansson, S. Thórarinsson & Th. Einarsson. C^{14}- aldursákvarðanir á sýnishornum varðandi íslenska Kvarterjarðfræði (English summary: C^{14} datings of Quaternary deposits in Iceland). *Náttúru-fræðingurinn 34*, 127-133.

Einarsson, Th. 1967: Zu der Ausdehnung der Weichselzeitlichen Vereisung Nordislands. *Sonderveröffentlichungen des Geologischen Institutes der Universität Köln 13*. 167-173.

Einarsson, Th. 1968: *Jarðfræði. Saga bergs og lands*. Mál og Menning, Reykjavík. 335 pp.

Einarsson, Th. 1973: Geology of Iceland. *In* M.G. Pitcher (ed.), *Arctic Geology. American Association of Petroleum Geologists Memoir 19*, 171-175.

Einarsson, Th. 1978: *Jarðfræði*. Mál og Menning, Reykjavík. 240 pp.

Einarsson, Th. 1979: On the deglaciation of Iceland. Abstract, 13th Nordic Geological Winter Meeting, *Norsk Geologisk Förening, Geolognytt 13*, 18.

Einarsson, Th. & K.J. Albertsson 1988: The glacial history of Iceland during the past three million years. *Philosophical Transactions of the Royal Society of London B 318*, 637-644.

Eyles, N. C.N. Eyles & A.D. Miall 1983: Lithofacies types and vertical profile models, an alternative approach to the description and environmental interpretation of glacial diamict and diamictite sequences. *Sedimentology 30*, 393-410.

Hjartarson, Á. 1989: The ages of the Fossvogur Layers and the Álftanes end-moraine, SW-Iceland. *Jökull 39*, 21-31.

Hjartarson, Á. & Ó. Ingólfsson 1988: Preboreal glaciation of southern Iceland. *Jökull 38*, 1-16.

Hjartarson, Á., F. Sigurðsson & Th. Hafstað 1981: *Vatnabúskapur Austurlands III. Lokaskýrsla*. Orkustofnun, Reykjavík. 198 pp.

Håkansson, S. 1983: A reservoir age for the coastal waters of Iceland. *Geologiska Föreningens i Stockholm Förhandlingar 105*, 65-68.

Ingólfsson, Ó. 1985: Late Weichselian glacial geology of the lower Borgarfjörður region, western Iceland: a preliminary report. *Arctic 38*, 210-213.

Ingólfsson, Ó. 1987: The Late Weichselian glacial geology of the Melabakkar-Ásbakkar cliffs Borgarfjörður, W-Iceland. *Jökull 37*, 57-81.

Ingólfsson, Ó. 1988: Glacial history of the lower Borgarfjörður region, western Iceland. *Geologiska Föreninges i Stockholm Förhandlingar 110*, 293-309.

Jóhannesson, H. & K. Sæmundsson 1989: Geological map of Iceland. 1:500.000. Bedrock geology. *Icelandic Museum of Natural History and Iceland Geodetic Survey, Reykjavík.*

Jux, U. 1960: Zur Geologie des Vopnafjord-Gebietes in Nordost-Island. *Geologie 9*. 1-57.

Mangerud, J., S.T. Andersen, B.E. Berglund & J.J. Donner 1974: Quaternary stratigraphy of Norden, a proposal for terminology and classification. *Boreas 3*, 109-127.

Miall, A.D. 1977: A review of the braided river depositional environment. *Earth-Science Reviews 13*, 1-62.

Miall, A.D. 1978: Lithofacies types and vertical profile models in braided river deposits: a summary. *In* A.D. Miall (ed), *Fluvial sedimentology. Canadian Society of Petroleum Geologists, Mem 5*, 597-604.

Norðdahl, H. 1983: Late Quaternary stratigraphy of Fnjóskadalur, Central North Iceland, a study of sediments, ice-lake, strandlines, glacial isostasy and ice-free areas. *LUNDQUA Thesis, 12*. Department of Quaternary Geology, University of Lund 78 pp.

Norðdahl, H. & Th. Einarsson 1988: Hörfun jökla og sjávarstöðubreytingar í ísaldarlok á Austfjörðum. *Náttúrufræðingurinn 58*, 59-80.

Norðdahl, H. & H. Hafliðason 1990: Skógar tefran, en sen glacial kronostratigrafisk markera på Nordisland. Abstract, 19th Nordic Geological Winter Meeting, *Norsk Geologisk Forening, Geonytt 17*, 84.

Norðdahl, H. & C. Hjort 1987: Aldur jökulhörfunar i Vopnafirði. In, *Ísaldarlok á Íslandi*. Abstract volume, Jarðfræðafélag Íslands, Reykjavík. 18-19.

Ólafsdóttir, Th. 1975: Jökulgarðar á sjávarbotni út af Breiðafirði (English summary: A moraine ridge on the Iceland shelf, west of Breiðafjörður). *Náttúrufræðingurinn 45*, 31-36.

Pétursson, H. 1986: *Kvartergeologiske undersøkelser pa Vest-Melrakkaslétta, Nordøst Island*. Unpublished Cand. Real Thesis, University of Tromsø, Tromsø. 157 pp.

Sæmundsson, K. 1974: Evolution of the axial rifting zone in northern Iceland and the Tjörnes fracture zone. *Geological Society of America Bulletin 85*, 495-504.

Sæmundsson, K. 1977: Geological map of Iceland, sheet 7, NE-Iceland. *Museum of*

Natural History and the Icelandic Geodetic Survey, Reykjavík.

Sigbjarnarson, G. 1983: The Quaternary Alpine glaciation and marine erosion in Iceland. *Jökull 33*, 87-98.

Steindórsson, S. 1962: On the age and immigration of the Icelandic flora. *Societas Scientiarum Islandica.* Rit XXXV, 157 pp.

Thórarinsson, S. 1937: The main geological and topographical features of Iceland. *Geografiska Annaler 19*, 161-175.

Thoroddsen, Th. 1904: Þættir úr jarðfræði Íslands. *Andvari XVIV*, 2-62.

Thoroddsen, Th. 1905-06: *Island, Grundriss der Geographie und Geologie.* Petermanns Mitteilungen, Ergänzungsheft no. 152 und 153. Justus Perthes, Gotha, 358 pp.

Thoroddsen, Th. 1911: *Lýsing Íslands 2.* Hið Íslenska Bókmenntafélag, Köpenhamn, 673 pp.

Formation and Deformation of Glacial Deposits, Warren & Croot (eds) © 1994 Balkema, Rotterdam, ISBN 90 5410 096 6

Till genesis and deglaciation in the Värnamo area, central southern Sweden

Lennart Sorby
Department of Quaternary Geology, University of Lund, Sweden

ABSTRACT: The town of Värnamo is situated in the Lagaån river valley. The river valley, trending north-south is a large fissure valley associated with the Protogine zone, the major fault zone through central southern Sweden. The valley is surrounded by an undulating plateau, which has been suggested to be a part of a Tertiary peneplain. A bedrock dependent morphology prevails on the plateau. A thin lodgement till discontinuously covers the bedrock. This is overlain by a till bed which mainly consists of sediment gravity flows intrabedded with meltwater deposits. The sediments commonly exhibit load and water-escape structures. The valley is filled with a thick pile of glaciolacustrine sediments. The deglaciation took place in a stagnant ice environment with melt-out and sediment gravity flow processes on the plateau, while calving and subaquatic deposition prevailed in the valley. The Glacial Lake Vidöstern had a much more restricted extent than has been previously postulated.

1. INTRODUCTION

The deglaciation of this area (Fig. 1) and its surroundings have so far not been studied in depth. The themes most discussed are the deglaciation pattern, deglaciation chronology and extent of glacial lakes.

Recent studies to the north and south of the investigated area (Bjelm 1976, Waldemarson 1986, Möller 1987) and the new Quaternary maps (Daniel 1986, 1989, Fredén 1988) suggest that the stratigraphy, deglaciation and extent of glacial lakes have a more varied pattern than previously postulated.

Nilsson (1937, 1942, 1953, 1958, 1968) and Rydström (1971) suggested large glacial lakes in the interior of southern Sweden during the deglaciation. However, Nilsson's and Rydström`s interpretations have been criticized by Bjelm (1976) and Waldemarson (1986) who suggested that the previously presumed glaciolacustrine sediments were deposited subaerially. According to Möller (1987) the isostatic rebound has a lower gradient across the isobases than pre-viously assumed, implying much more restricted glacial lakes.

Areas with hummocky moraine and cover moraine, i e "a veneer of till, the surface of which consists of a low relief dominantly devoid of transverse or linear elements" (Aario 1977), in the Växjö area (Fig. 1), have been interpreted by Möller (1987) to have been deposited as a complex of subglacial melt-out till and supra-glacial melt-out till. Daniel (1986, 1989) and Fredén (1988) have mapped several areas with hummocky moraine between areas of cover moraine and streamlined terrain. Hummocky moraines are associated with areas of thicker Quaternary deposits or areas with glaciofluvial systems. Rydström (1971) interpreted sorted sediments covered by till as "oscillation deposits". Oscillation deposits were found in hills with drumlinized form, mainly in the Växjö area but also in the Värnamo area.

The deglaciation pattern suggests deglaciation with stagnant ice south of Växjö (Björck & Möller 1987, Möller 1987) and two, probably late-glacial, readvances in the Vättern basin in

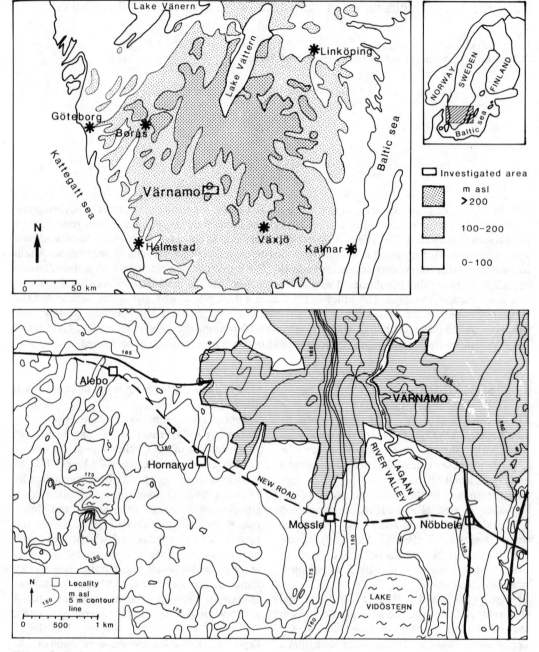

Fig. 1. Location maps.

the north (Bjelm 1976, Waldemarson 1986) (Fig. 1). No detailed investigations have been presented from the area in between, but it has been assumed that the ice-sheet margin coincided with the activity limit during the deglaciation (Lagerlund et al. 1983), i e there was no stagnant ice in front of the glacier.

The deglaciation chronology through central

southern Sweden was first established by Nilsson (1968), based on varve chronology. It has been revised by Berglund (1979), Lagerlund et al. (1983) and Björck et al. (1988), based on correlations between ^{14}C ages and varve chronology along the east coast of southern Sweden. A rapid deglaciation, approximately 300 m/y, between 12 350 - 12 050 BP is suggested for the region (Björck et al. 1988).

The Värnamo area is in the center of this region. A new road through the area with several exposures has provided new information.

2. GEOLOGICAL SETTING

The Värnamo area (Fig. 1) is situated in central southern Sweden, 140 - 180 m asl, and above the marine limit/highest coastline. There are several lakes in the area, trending north-south, and associated with bedrock structures. The area is situated along the north-south trending Protogine zone, the major fault zone through southern Sweden. The bedrock mainly consists of gneiss west of the zone and granite to the east.

The morphology of the pre-Quaternary landscape has been outlined by Mattsson (1962) and Lidmar-Bergström (1982, 1988). According to Lidmar-Bergström (1988) the area is situated on the border between a Tertiary peneplain in the south and a Cretaceous denudation area in the north, coinciding with areas below and above 200 m asl, respectively (Fig. 1). The peneplain has a relief of 0-25 m and includes a few residual hills (inselbergs), while the northern area includes several residual hills and has a higher relief, 25-50 m.

The Värnamo area consists of an undulating plateau at 170-180 m asl, which is part of a residual hill. The plateau is cut through by the north-south trending Lagaån river valley (Fig. 1). The surface of the valley sediments lies 30-40 m lower than the plateau with minor terraces on the slopes. The bedrock valley is approximately 120 - 130 m deep, with 90 - 100 m of sediment in the present valley bottom. The valley contains mainly fine grained sediments, while the plateau is covered by till. It has been suggested that glacial lakes have covered almost the whole area (Nilsson 1968) to an altitude of 176 m asl at Värnamo.

3. LOCALITIES ON THE PLATEAU

On the plateau, west of the Lagaån river valley, the undulating to hummocky landscape is due to bedrock morphology. The rather thin discontinuous till cover, normally is less than 3 m thick. However, in bedrock depressions it may exceed 5 m. The sediment stratigraphy and interpreted genesis are demonstrated in three exposures on the plateau.

3.1 Hornaryd.

The site is situated southwest of the town of Värnamo (Fig. 1). An exposure has been examined at a construction site (Figs. 2,3,4).

On the northward sloping smooth rock surface (gneiss) there is a small patch, about 1 m long and less than 4 cm thick, of dense compact homogenous silty diamicton, unit 1 (Fig 4). The fine grained material contains angular sand-sized fragments of the bedrock.

The unit is interpreted as a lodgement till. Silt-size particles are typical of glacier abrasion in the zone of traction (Drewry 1986, Dreimanis 1988). The dense compact material has been under high pressure, probably plastered on to the bedrock. The material is probably abrasional wear debris from the bedrock. The unit was probably deposited by ice moving from the north, ± 10^0, indicated by striae on stoss-sides in the area. The limited extent of the unit is probably due to deposition on the sloping surface rather than to later erosion, since the diamicton is so compact and hard it is unlikely that it could have been eroded by slow meltwater flow (see unit 2).

Unit 2, draping the till (unit 1) and the rock surface with a sharp contact, is 5 - 10 cm thick. The unit follows the bedrock surface from the lower steep northern side (Fig. 4) and disappears close to the ground surface in the south. It is thickest in the center of the exposure. It consists of a slightly laminated poorly sorted sediment. The laminae are dominated either by fine sand or silt. The unit includes several deformation structures which may be described as load-cast or water-escape structures.

An interpretation of the unit is hard to make because of the character and deformation

Fig. 2. An overview of the investigated section at Hornaryd showing the diamict sediments above the sloping bedrock. Marked area is shown in Fig. 3. The stick for scale is 2 m.

structures. However, the poorly sorted sediment was probably deposited down the sloping rock surface, with the highest accumulation where the slope is less steep (Figs. 2 & 4). The fine grained composition, similar to unit 1 and with no coarse clasts, may suggest that the source material was the same as in unit 1, the lodgement till. The fine grained material may, therefore, be melt-out material from the base of the glacier. A possible interpretation is that the sediments were basally released and deposited by gravity flows or by sheet flow at the ice/rock interface.

Unit 3 (Fig. 4) is composed of loose sandy silty diamicton with abundant pebbles and cobbles and a few boulders. It also contains many small (1 - 2 cm), rounded, massive intraclasts of sand. The clasts and the intraclasts occur scattered in the exposure, although the boulders seem to be concentrated in the lower part of the unit. Around the big boulders there are sandy and silty

laminae, often disrupted and mixed or interfingering with diamicton. The laminae show load structures under and almost surrounding the boulders. Disrupted sand/silt laminae of unit 2 intrude subhorizontally or vertically into unit 3.

The deformation is interpreted to be water-escape structures formed during compaction of the diamicton. The large boulders have caused the injection of sand/silt laminae of unit 2 into unit 3 by loading of water-soaked sediment during deposition. The sediment composition of unit 3 differs markedly from unit 1 and 2 and suggests a material derived from both plucking / crushing of coarse material and abrasional wear debris. The intraclasts may originally have been laminae of sand, but probably have been fragmented by flow processes and incorporated in the diamicton as intraclasts.

The sand and silt laminae, intraclasts, load and dewatering structures and loose character of unit 3 suggest a supraglacial sediment gravity flow

Fig. 3. Close up photo of the Hornaryd section (see Fig. 2).

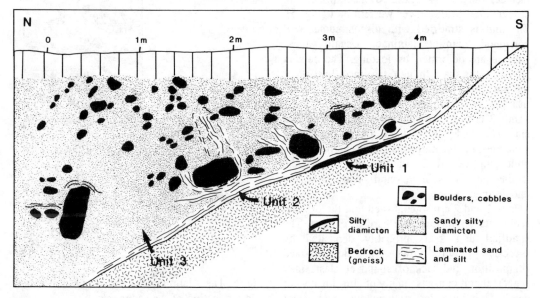

Fig. 4. Drawing of the section at Hornaryd (Fig. 2) showing the load and water-escape structures.

origin. The total dominance of diamicton facies and absence of glaciolacustrine sediments suggest a subaerial environment. During the supraglacial melting and formation of unit 3, stagnant ice was present between units 2 and 3. During the final melt-out, the unit 3 sediments were dumped on unit 2, resulting in water-escape and load structures. Deposition of units 2 and 3 most likely occurred in a zone of stagnant or dead-ice where the possibilities for unit 2 to remain unchanged are greatest (Drewry 1986).

3.2 Alebo.

In the western part of the area (Fig. 1), the new road cuts through two drumlinized bedrock hummocks. In a leeside position the bedrock is covered by a thin, 2 - 5 cm thick, dense, compact silty diamicton which appears plastered on the rock surface. The similarity with unit 1 at Hornaryd, interpreted as lodgement till, is obvious. Beneath the till there are fine striae indicating ice moving from the north.

Between the drumlinized hummocks is a bedrock depression. It is filled up with sediments which have been classified into three units (Fig. 5).

Unit 1 consists of sandy, silty diamicton. It includes only a few scattered clasts, which are concentrated in the lower western part (Fig. 5). The unit is stratified with discontinuous sub-horizontal diamict laminae. Usually these laminae are deformed by loading. The contact with unit 2 is sharp though deformed. In parts of the unit there are small (1 - 10 cm) rounded, massive, sandy intraclasts.

Unit 2 consists of laminated sorted sand and silt. The beds show load and water-escape structures, but the unit usually has sharp contacts with unit 1 and 3. In the upper part the lamination is subhorizontal and relatively undeformed.

Units 1 and 2 are so contorted that the original sedimentary structures are hard to discern. The stratified unit 1 is probably deposited by succesive gravity flows. The fine grained composition and lack of abundant clasts may suggest that it consists mainly of abrasional wear debris. Unit 2 was deposited by minor melt-water

flows into the bedrock depression. The deformation structures are probably due to loading of the liquified sediments during sedimentation. The deformation could have been enhanced by melting of buried iceblocks beneath.

Unit 3 covers the lower units and is the uppermost unit in the profile. It consists of different beds of diamictons with sandy, silty composition. The unit has few large clasts but includes some scattered pebbles and cobbles. Some beds are stratified. Sandy and silty intraclasts occur in the lower part of the unit and are about 10 - 20 cm (across), with weak lamination. Several thin subhorizontal beds of reddish diamicton and also some intraclasts of sand occur in the unit. The reddish diamicton has the same composition as the surrounding material, but the colour is hard to explain.

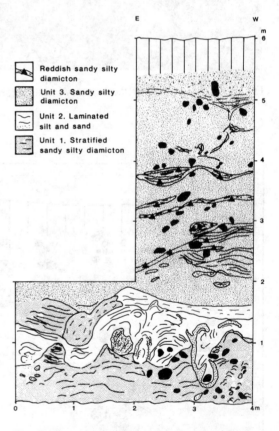

Reddish sandy silty diamicton

Unit 3. Sandy silty diamicton

Unit 2. Laminated silt and sand

Unit 1. Stratified sandy silty diamicton

Fig. 5. The Alebo section showing stratification and lamination of the sediments.

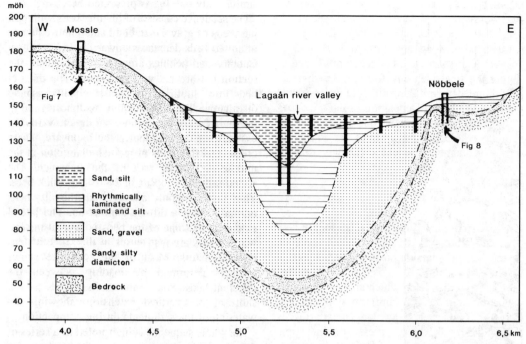

Fig. 6. Schematic section through the Lagaån river valley based on geotechnical corings (vertical black lines) and investigations at Mossle and Nöbbele.

The material is in general coarser than in the lower units and contains more clasts. The stratification and beds of diamicton with slight differences in texture and colour suggest that unit 3 was deposited as series of succesive debris flows. The source material of the reddish diamicton is unknown but probably related to the fine grained fractions since the petrographical composition is the same (crystalline bedrock) as in the surrounding material.

A thin basal till drapes the bedrock down on the leeside of the drumlin, suggesting that the ice was probably active in the depressions as well. The stratified diamictons and differences in texture, composition and colour, suggest that both unit 1 and 3 sediments were deposited as succesive gravity flows, interbedded with sandy material reflecting a minor influx of meltwater. Unit 1 and 2 sediments show deformation structures, probably related to loading of liquified sediments. The deformation could have been enhanched by buried dead-ice beneath. It is possible that units 1 and 2 were deposited on

dead-ice which remained in the bedrock depression. Unit 3 which is relatively undeformed was probably deposited by sediment gravity flow after melting of the buried ice. Melting of stagnant ice probably generated the sediment gravity flows. During deglaciation deposition of units 1, 2 and 3 is most likely to have occurred in a stagnant ice environment.

3.3 Mossle.

A rather flat arable area on the plateau, to the west of the Lagaån river valley, passes into the rather steep western valley side (Fig. 1). On the plateau in general the till cover is thin, 1 - 2 m, on the undulating bedrock. Thick sediments have accumulated on the upper part of the valley side and form a 200 m broad extention of the plateau around Mossle (Fig. 6). The new road cuts through these thick Quaternary deposits. The exposure has been examined in three different sections. However, since the stratigraphy is

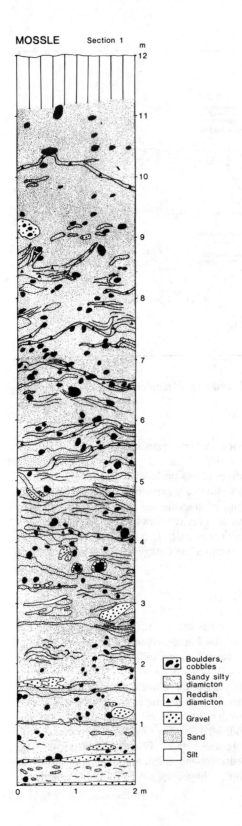

MOSSLE   Section 1

Legend:
- Boulders, cobbles
- Sandy silty diamicton
- Reddish diamicton
- Gravel
- Sand
- Silt

similar only one log is presented here (Fig. 7).

The sequence consists of diamict beds with thin intrabeds of gravel, sand and silt. The frequency of sorted beds decreases upward in the sections. Cobbles and pebbles are spread throughout the section but are not very abundant. The diamict beds are stratified or massive silty, sandy diamictons. The stratification is subhorizontal in the main part of the section, but dips towards the east further down-slope in the exposure. There are several thin beds of reddish diamicton in the middle and upper part of the sequence. The sorted sediments occur in 2 - 20 cm thick beds. They are commonly massive, but the silty beds normally have a diffuse lamination. The lateral extension of some of the beds is several meters, while others are just lenses in the range of tens of centimeters. Most of the sequence shows contacts deformed by loading between the different beds. Some sand and silt beds have a short almost vertical extension, showing dewatering of the sediments during compaction.

The whole sequence is interpreted as a series of diamict beds that were deposited as gravity flows, alternating with influx of sorted sediments deposited by meltwater flows. The stratification suggests that deposition took place on a subhorizontal bed.

Compared to the thin sediment cover on the flat plateau, the accumulation on the valley side is remarkable (Fig. 6). The thick sequence of material suggests that there must have been a rich source of debris in the area. Down-hill processes with material from the plateau, could, at least in the lower part of the sequence, be a supplier of material. However, the subhorizontal bedding does not support this origin, and further up in the sequence it is unlikely since the inclination towards the valley is too low to allow sediment gravity flow. This rather suggests debris-rich active or dead ice as the supplier of material.

The accumulation is located on the valley side with a sediment thickness up to 12 m. The sediment cover on the plateau is normally thin

Fig. 7. Profile through the Mossle section. The log is representative for the character of the deposits.

and usually depends on the bedrock morphology. On the valley side the bedrock surface dips down into the valley while the sediment thickness increases and the surface has approximately the same level as the plateau. A possible explanation is that the thick sequence is built up as a lateral moraine to an ice body in the Lagaån river valley. If no ice body existed in the valley, a thin sequence following the bedrock topography would be expected. The lateral ice-contact could be a zone of ice-cored lateral moraines, which is common along recent glaciers. During melting the ice-cored moraine generated both the gravity flows and meltwater flows in the area.

4. THE LAGAÅN RIVER VALLEY

The valley is a bedrock trench approximately 130 m deep from the plateau surface, filled with up to 90 m of sediment. Geotechnical surveys and exposures on the eastern side of the river valley have provided a schematic section across the valley (Fig. 6).

On the eastern side of the river valley there is a terrace approximately 5 m above the valley floor. The terrace consists of sediment-covered bedrock. A hummocky bedrock surface is visible in several bedrock knobs in the terrace. Till, probably covering bedrock, has also been mapped on the terrace. The terrace sloping towards the valley floor, probably marks the transition to the true bedrock trench.

At Nöbbele, on the terrace, several profiles in a large road cut have been investigated. The stratigraphy, as traced along the exposures, is summarized in a log (Fig. 8).

In the road profile a bedrock knob is exposed, covered by a 20 - 200 cm thick gravelly sandy diamicton with weakly developed stratification. It is rich in pebbles and cobbles. There are several thin lenses of sand and silt in the diamicton. Above the diamicton with a sharp contact is a unit of tabular laminated sand and silt. A unit of trough-cross laminated sand follows above with a gradational contact. Paleocurrent is towards S or SW, i e along the terrace or towards the valley.

There is a gradual transition to the next unit, which consists of sand, silt and clay in fining upward sequences. The sand usually has current

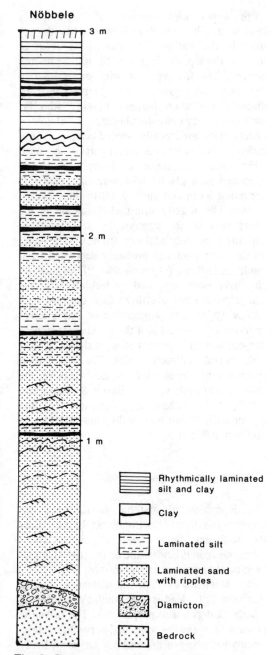

Fig. 8. Composite log for the Nöbbele section.

ripples draped by undulating silt laminae grading into more or less horizontal laminated clay. The silt/clay laminae show different types of load structures or convolute bedding in the lower part of the unit.

The upper unit consists of rhythmically laminated silt and clay. The unit is thicker towards the valley. The lamination is finer towards the top, grading into the soil weathered horizon. The fine upper lamination is hard to discern and gives the impression of discontinuous short laminae of silt and clay. Synsedimentary convolute bedding, overlain and underlain by horizontally bedded layers, occur at different levels and in different parts of the unit.

The whole sequence is interpreted to be deposited in a glacial lake with the sediments becoming more and more distal to the receding glacier. The slightly stratified diamicton in the lower part of the stratigraphy was probably deposited by debris flows from the receding glacier. The sand was probably deposited from turbid underflows (cf Gustavson 1975). Silt and clay have been deposited by turbidity currents and rain-out from overflows into the basin (cf Ashley 1975). The large influx of fine grained material has resulted in a thick sequence, where rain-out was an important process shown by clay beds several centimeters thick. The deformation structures are interpreted as load-casts in the lower sandy units (e g Collinson & Thompson 1982) and subaqueous slumping in the rhythmically laminated sediments (Leckie & McCann 1980).

5. DISCUSSION

The deposition of material in the area has occurred in different environments. In the valley, the sediments have been deposited under subaquatic conditions, while on the plateau under subglacial and subaerial conditions.

The sediment cover on the plateau is usually thin with little dead-ice morphology, leading to a bedrock dependent morphology. The reason is probably that the ice contained very little debris. Deep weathering of the bedrock has most likely been the main influence in the evolution of the bedrock morphology (Lidmar-Bergström 1988) which is just modified by glacial activity. Abrasion of the rather smooth substrate probably did not resulted in large volumes of debris being picked up by the glacier. This is demonstrated in the area by the relative scarcity of coarse clasts, fine grained composition of

sediments and thin till deposits covering the area, and also by the few localities with roche moutonnées or plucked surfaces. Even in the fractured zones the angular boulders seem not to have been more than slightly crushed or displaced a couple of meters from their original position. The limited erosion of the substrate has also been noted by Knutsson (1962) and Rydström (1971) from the Växjö area, and observations of pre-Quaternary weathering material from the same area (Knutsson 1962).

A compact fine grained diamicton, interpreted as a lodgement till, occurs discontinuously in the area. The lodgement till consists of mainly silt and sand with angular sand particles of the bedrock indicating that the sediment is abrasional wear debris. Below the lodgement till there are fine striations reflecting an ice movement from the north. The striation probably marks the last ice-movement direction since the striation is the same even where the till is absent. This ice-movement probably deposited the lodgement till in the area. The limited extent of lodgement till at the Hornaryd locality may suggest that deposition on abraded smooth bedrock is related to the inclination of the rock surface.

The sediments at the Hornaryd, Alebo and Mossle localities show that the deglaciation on the plateau was characterized by supraglacial melt-out and sedimentation by gravity flows (Fig. 9). The resulting sediments were probably derived from stagnant or dead ice in front of an active ice-margin. The dead-ice zone was probably not more than a few hundred meters, since the deglaciation of the region was rather fast in the ameliorated climate during the Bölling chronozone (Björck et al. 1988) and the frontal thermal conditions in general, have been postulated to be warm-based (Lagerlund et al. 1983).

The large sediment accumulation at Mossle on the upper valley side indicates that the valley was filled with ice during the deglaciation. The Lagaån river valley / Lake Vidöstern is about 2 km wide and 150 m deep. The ice body in the valley producing the lateral moraine would therefore have been more than 150 m thick. An ice body more than 150 m thick, standing in a glacial lake with a water depth of 100 m, will most likely be an active glacier with a calving ice-margin. The thick deposits on the valley side,

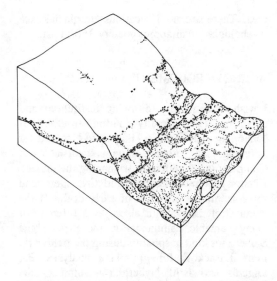

Fig. 9. Schematic drawing of the dead-ice melting and deglaciation of the Värnamo area on the plateaus.

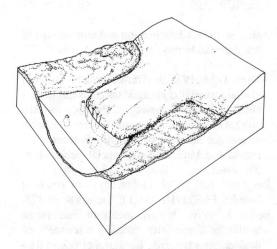

Fig. 10. Schematic drawing of the deglaciation of the Lagaån river valley with a minor ice-lobe in Glacial Lake Vidöstern with a calving front. The thick deposits at Mossle were deposited in a position lateral to the ice-lobe.

ice thickness and water depth suggests a model with a minor ice-lobe standing in a glacial lake rather than a disintegrating stagnant ice or an ice-margin with an embayment due to rapid calving (Fig. 10). The glacier was probably channeled

into the bedrock valley by the surrounding hills. The limited depth of the valley suggests that it was only during the deglaciation with a quite thin glacier that the valley affected the glacial pattern. Lateral moraines on a valley side related to an ice-lobe have also been described by Hilldén (1984) from the Borås area (Fig. 1). Ice-lobes have also existed or readvanced in other valleys in central or western south Sweden during the deglaciation (Waldemarson 1986, Ronert 1991). It is possible that minor ice-lobes existed in several other deep elongated lakes in this region during the deglaciation. On the other hand, calving in local glacial lakes may have resulted in embayments of the ice-front (Daniel 1986).

Sorted sediments covered by till, "oscillation deposits" (Rydström, 1971), in hills with a drumlinized form have been reported from the Växjö area (Rydström 1971, Daniel 1989) and the Värnamo area (Rydström 1971). The sediments often occur along valley sides and were assumed to be older deposits overridden by the Late Weichselian ice advance. The area west of the Lagaån river valley, with the Hornaryd, Alebo and Mossle localities, has a large-scale drumlinized morphology. Observations in the northern part of the area, show glaciofluvial sand and gravel below till with glaciotectonic structures in the glaciofluvium (Rydström 1971). They were therefore interpreted as "oscillation deposits" (Rydström 1971). In the exposures at Hornaryd, Alebo and Mossle there are, however, no observations suggesting either that the sediments have been overridden by ice or that they are older than the last deglaciation.

The geotechnical survey gives a schematic view of the sediments in the Lagaån river valley trench (Fig. 6). The fining upward sequence of bottomset beds indicates a receding glacier with increasingly distal sedimentation. The terrace on the eastern side is probably related to bedrock knobs on the valley side. The sedimentary sequence in the sections at the Nöbbele locality is the same as have been noted during the geotechnical surveys. The Nöbbele section is, therefore, a sedimentological mirror of the deep trench. The sediments at Nöbbele are more coarse grained relative to those in the valley bottom. This may be due to lesser water depth at Nöbbele and a position closer to the eastern plateau. The altitude of the fine grained terrace surface is slightly below 155 m asl. No observations in the area suggest any higher levels than 160 m asl for Glacial Lake Vidöstern during the deglaciation. This rejects the model postulated by Nilsson (1968).

CONCLUSION

The deglaciation of the Värnamo region and the central part of southern Sweden is in general agreement with Lagerlund et al. (1983) and Björck et al. (1988), but is considered to be slightly more complex than earlier suggested.

A thin lodgement till is discontinuously spread on the rock surfaces. It consists of abrasional wear debris from the crystalline bedrock. The discontinuous deposition may be due to the smooth rock surface and the inclination of the rock surface.

The lodgement till represents the last ice-movement, shown by the combination of striations from the north below the thin massive till, and the lack of other striation directions.

The main part of the total sediment cover has been deposited by sediment gravity flow in a supraglacial position.

Reddish diamicton occurs in some exposures. The origin is unknown but if traced, perhaps it could act as a marker horizon in stratigraphy and indicator of ice-movement direction.

The general deglaciation pattern in the investigated area has probably been characterized by a marginal zone of stagnant or dead ice. The ameliorated climate during the deglaciation, Bölling chronozone, caused a rapid disintegration of the stagnant zone.

The smooth rock surfaces with a fine striation, the fine grained composition of tills, thin till cover, low frequency of roche moutonnées and coarse clasts and poorly eroded fracture zones, indicate that glacial abrasion, plucking and freezing on has been low.

In the Lagaån river valley a minor ice-lobe, calving in Glacial Lake Vidöstern, seems to have existed during the deglaciation.

Turbid underflows and rain-out of fine grained material are the major agents for depositing the Lagaån river valley sediments in the investigated

area. There are no indications of glacial lake levels higher than approximately 160 m asl.

ACKNOWLEDGEMENT

I wish to thank the following institutions and persons: The County Administration of Jönköpings län and especially Eric Haglund for initiating the project. The Swedish Road Company in Jönköping for financing the project. Kärstin Malmberg-Persson and Erik Lagerlund for constructive criticism of early drafts of the manuscript. David Mickelson, as a referee, for many valuable comments on the paper. Inga Sjöberg for kind hospitality during the fieldwork. Lena Barnekow for grain-size analyses. Per Lagerås and Britt Nyberg for some of the drawings and Thorsteinn Saemundsson for help with the computer.

REFERENCES

Aario, R. 1977. Classification and terminology of morainic landforms in Finland. Boreas 6: 87-100.

Ashley, G.M. 1975. Rhythmic sedimentation in glacial Lake Hitchcock, Massachusetts-C o n - necticut. In: A.V. Jopling & B.C. McDonald (eds), Glacifluvial and glacilacustrine sedi- mentation. Society of Economic Paleonto- logists and Mineralogists, special publications 23: 304-320.

Berglund, B.E. 1979. Deglaciation of southern Sweden 13.500 -10.000 B.P. Boreas 8: 89-118.

Bjelm, L. 1976. Deglaciationen av Småländska höglandet, speciellt med avseende på deglaciationsdynamik, ismäktighet och t i d s - ställning. LUNDQUA, Thesis 2, Department of Quaternary Geology, University o f L u n d, Sweden. 78 pp.

Björck, S. & P. Möller, 1987. Late Weichselian Environmental History in Southeastern Sweden during the deglaciation of the Scandinavian Ice Sheet. Quaternary Research 28: 1-37.

Björck, S., B.E. Berglund & G. Digerfeldt, 1988. New aspects on the deglaciation chronology of south Sweden. Geographia polonica 55: 37-49.

Collinson, J.D. & D.B. Thompson, 1982.

Sedimentary structures. George Allen & Unwin, London. 194 pp.

Daniel, E. 1986. Beskrivning till Jordartskartan Värnamo SO. Sveriges Geologiska Undersökning, Ae 80. 60 pp.

Daniel, E. 1989. Beskrivning till Jordartskartan Växjö SV. Sveriges Geologiska Undersökning, Ae 101. 77 pp.

Dreimanis, A. 1988. Tills: Their genetic terminology and classification. In: R.P. Goldthwait & C.L. Matsch (eds), Genetic Classification of Glacigenic Deposits. 17-83. Balkema, Rotterdam.

Drewry, D. 1986. Glacial Geologic Processes. Arnold. 276 pp.

Fredén, C. 1988. Beskrivning till Jordartskartan Värnamo SV. Sveriges Geologiska Undersökning, Ae 93. 58 pp.

Gustavson, T.C. 1975. Sedimentation and physical limnology in proglacial Malaspina Lake, southeastern Alaska. In: A.V. Jopling & B.C. McDonald (eds), Glacifluvial and glacilacustrine sedimentation. Society of Economic Paleontologists and Mineralogists, special publications 23: 249-263.

Hilldén, A. 1984. Beskrivning till jordartskartan Borås SO. Sveriges Geologiska Undersökning, Ae 58. 65 pp.

Knutsson, G. 1962: Algutsbodatraktens geologi. Algutsbodaboken, 3-48. Nybro.

Lagerlund, E., G. Knutsson, M. Åmark, M. Hebrand, L-O. Jönsson, B. Karlgren, J. Kristiansson, P. Möller, J.M. Robison, P. Sandgren, T. Terne & D. Waldemarson, 1983. The deglaciation pattern and dynamics in south Sweden. LUNDQUA, Report 24. Dept of Quaternary Geology, University of Lund. 7 pp.

Leckie, D.A. & S.B. McCann, 1980. Glacio-lacustrine sedimentation on low slope p r o - grading delta. In: R. Davidson-Arnott, W. Nickling, & B.D. Fahey, (eds), Research in glacial, glacio-fluvial and glacio-lacustrine systems. 261-278. 6th Guelph symposium on Geomorphology, Proceedings. Geobooks, Norwich, Connecticut.

Lidmar-Bergström, K. 1982. Pre-Quaternary geomorphological evolution in southern Sweden. Sveriges Geologiska Undersökning C 785. 202pp.

Lidmar-Bergström, K. 1988. Denudation surfaces of a shield area in south Sweden. Geografiska Annaler 70A(4): 337-350.

Mattsson, Å. 1962. Morphologische Studien in Südschweden und auf Bornholm über die nichtglaziale Formenwelt der Felsenskulptur. Lund Studies in Geography A 20, Department of Physical Geography, University of Lund, Sweden . 357 pp.

Möller, P. 1987. Moraine morphology, till genesis, and deglaciation pattern in the Åsnen area, south-central Småland, Sweden. LUNDQUA, Thesis 20, Department of Quaternary Geology, University of Lund, Sweden. 146 pp.

Nilsson, E. 1937. Bidrag till Vätterns och Bolmens senkvartära historia. Geologiska Föreningens i Stockholm Förhandlingar, 59: 189-204.

Nilsson, E. 1942. Gotiglaciala issjöar i södra Sverige. Geologiska Föreningens i Stockholm Förhandlingar, 64: 143-159.

Nilsson, E. 1953. Om Södra Sveriges senkvartära historia. Geologiska Föreningens i Stockholm Förhandlingar, 75: 155-246.

Nilsson, E. 1958. Issjöstudier i södra Sverige. Geologiska Föreningens i Stockholm Förhandlingar, 80: 166-185.

Nilsson, E. 1968. Södra Sveriges Senkvartära Historia. Geokronologi, issjöar och landhöjning. Kungliga Svenska Vetenskapsakademins Handlingar, Ser 4, Band 12, no 1. 117pp.

Ronert, L. 1991. The character of deglaciation in the Berghem area, southwestern Sweden. Geologiska Föreningens i Stockholm F ö r - handlingar, 113: 295-308.

Rydström, S. 1971. The Värend district during the last glaciation. Geologiska Föreningens i Stockholm Förhandlingar, 93: 537-552.

Waldemarson, D. 1986. Weichselian lithostratigraphy, depositional processes and deglaciation pattern in the southern Vättern basin, south Sweden. LUNDQUA, Thesis 17, Department of Quaternary Geology, University of Lund, Sweden. 128 pp.

Formation and Deformation of Glacial Deposits, Warren & Croot (eds) © 1994 Balkema, Rotterdam, ISBN 90 5410 096 6

The stratigraphy and sedimentary structures associated with complex subglacial thermal regimes at the southwestern margin of the Laurentide Ice Sheet, southern Alberta, Canada

David J.A. Evans
Department of Geography and Topographic Science, University of Glasgow, UK

ABSTRACT: Complex stratigraphic sequences in the cliffs of One Tree Creek, southern Alberta reveal seven lithofacies associations (LFA's) which have been disturbed by two generations of glacitectonics. LFA's 1 and 2 (preglacial fluvial and lacustrine sediments), LFA 3 (grey till) and LFA 4 (supraglacial sediments) were all disturbed by the first generation glacitectonic event which terminated when either ice or a deforming till layer truncated the contorted sediments. Above the erosional contact formed at the end of the first generation glacitectonic event, a lodgement till forms the base of LFA 5 which grades upwards into a gravitationally deformed melange of debris flow, fluvial and subaqueous origin (subglacial cavity fills). LFA 6 is a diamicton which records the glacial molding of pre-existing material followed by melt-out. A prominent erosional contact separating LFA's 5 and 6 suggests renewed contact of the glacier with its bed after a period of decoupling. Second generation glacitectonics acted to disturb LFA's 2, 3, 4 and 5 after renewed ice/till coupling which was a response to renewed fast glacier flow or surging.

The LFA's and glacitectonic structures in the One Tree Creek sections are related to sediment properties which were in turn related to the discontinuous permafrost and proglacial lakes overrun by the ice sheet and to changing subglacial thermal regimes. Subglacial cavity fills and the second generation glacitectonics may relate to periods when large amounts of meltwater were being generated regionally by the ice sheet. Many ice thrust terrains are cross-cut by subglacial meltwater channels, indicating that they date to initial ice advance. Shifting reservoirs of subglacial meltwater may have had profound effects on the distribution of both fluviglacially and glacially derived flutings. The initiation of one of the processes of subglacial tunnel valley cutting (piping), cavity filling, ice streaming/till deformation and glacier surging in one area would have brought about a chain reaction elsewhere.

INTRODUCTION

The role of thermal zones in the construction and deposition of glacigenic landforms and sediments at the margins of the Laurentide Ice Sheet has been stressed by numerous researchers (eg. Clayton and Moran 1974; Aber 1982, 1985; Bluemle and Clayton 1984; Clayton et al. 1985; Mooers 1990). Stratigraphic and morphologic evidence has also been presented for glacier surging which was responsible for severe glacitectonic disturbance (Clayton et al. 1985; Campbell and Evans 1990). Southern Alberta has long been recognised as an area of classic glacitectonic disturbance and many geomorphic and stratigraphic examples have been cited from there (Slater 1927; Kupsch 1962; Moran 1971; Stalker 1973a, 1976; Moran et al. 1980).

Although many stratigraphic

sequences have been presented for southern Alberta, quite often the more complex and disturbed exposures along the extensive cliffs of the main rivers have been overlooked. The purpose of this paper is to present the complexity of some Quaternary stratigraphic sequences in the southern Alberta plains, especially as continuous cliff exposures in contrast to composite logs. Extensive stratigraphic exposures in the One Tree Creek drainage basin, a tributary of the Red Deer River, provide important information on former glacitectonics and ice sheet dynamics in the area (Fig.1).

2 GENERAL STRATIGRAPHY

The general stratigraphy of the area is discussed by Evans and Campbell (1992) who identify seven main lithofacies associations (LFA's 1-7). These form predominantly tabular stratigraphic sequences that have been glacitectonically or gravitationally disturbed at some locations (Fig.2).

The LFA's can be used as a local stratigraphic succession for comparison with the disturbed sedimentary sequences in the One Tree Creek drainage basin. The most intense ductile folding and attenuation is displayed by the lower/older diamicton facies whereas thrust faulting has occurred in underlying sands and gravels. The latter is associated with both the folding of the older diamicton and with younger, final ice advances which disturbed most of the stratigraphic sequence at some locations (cf. Campbell and Evans 1990). It is suggested that at least two major glacitectonic events have occurred and that the migration of the frozen bed zone of the ice sheet margin has had profound effects on the nature of the sedimentary succession and glacitectonic disturbance. Gravitational folding and faulting has occurred within the proglacial lake sediments of immediate postglacial age.

The preglacial fluviatile sediments (LFA 1) outcrop only in section 11 of the One Tree Creek basin (Fig.2). LFA 2 (first proglacial lake sediments) occur within all sections and have undergone various degrees of disturbance. A dark grey diamicton (till; LFA 3) occurs in most sections and is the most heavily disturbed of all lithofacies. It contains intraclasts of LFA 2 which appear to have been ingested after the closure of overfolds in the till around rafts of LFA 2. Elongate rafts of LFA 2 may have

originated as interbeds with LFA 3 which are apparent in sections 3 and 6 (Fig.2). Supraglacial gravels, sands and diamictons (LFA 4) are poorly represented but are most prominent in sections 7 and 8 where they have been faulted and contorted in association with the deformation of LFA 3 (Figs. 2, 4-6). A heterogeneous diamicton (LFA 5) is interpreted as a subglacial cavity-fill deposit. The massive structure and strong NNE-SSW clast fabric of the basal part of this LFA suggest deposition by lodgement but this gave way to debris flow, fluvial and subaqueous deposition during a period when the ice/substrate boundary was characterised by large cavities (Figs.4-7). An erosional contact separates LFA's 1-5 from the overlying diamicton (LFA 6) which is interpreted as a lodgement, deformation or melt-out till depending upon local characteristics (Fig.8). LFA 6 records a renewed contact by the glacier with its bed, bringing erosion of LFA's 1-5 and their minor glacitectonic disturbance. Shear structures and clast fabrics in LFA 6 appear to record several ice movement directions possibly due to glacier surging brought on by meltwater decoupling and/or movement over a deforming bed of LFA 6.

Postglacial sediments deposited in lacustrine, fluvial and aeolian environments are represented by LFA's 7a-e. In the context of this paper LFA 7a is the most important of the postglacial deposits. These are proglacial lacustrine sediments relating to the damming of lakes by retreating ice and include extensive interbedded sediment mass flows, derived from reworked LFA 6, and gravitationally disturbed rhythmites suspended in stratified diamictons (Fig.9).

3 FIRST GENERATION GLACITECTONICS ASSOCIATED WITH LFA'S 2-4

Although the exact ages of depositional events are unknown, the deposition of LFA 2 in a proglacial lake was followed by the deposition of LFA 3 during a period of glaciation. In the One Tree Creek basin, LFA 2 fines upwards from massive and micro-laminated sandy silts to silt/clay rhythmites. As LFA 3 has been severely disturbed, its exact origin at some locations is uncertain but Evans and Campbell (1992) consider it collectively to have the characteristics of all till types. These have undergone postdepositional deformation thus explaining the contorted LFA 2 intraclasts or melt-out sand lenses that occur in LFA

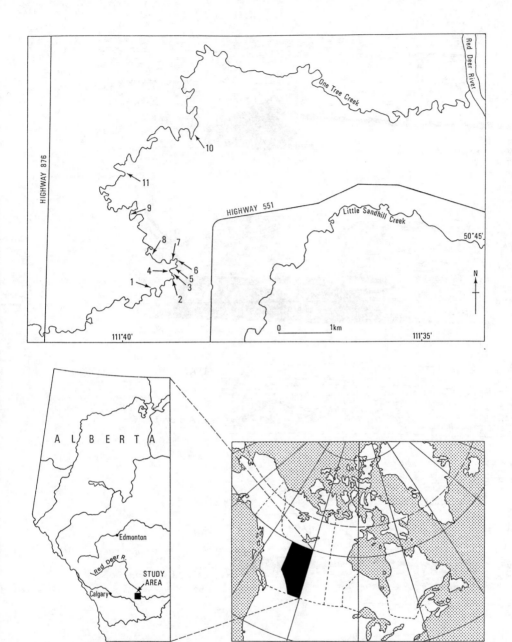

Figure 1: Location map of Alberta and One Tree Creek. Sections OTC 1-11 are marked 1-11.

3 which were probably rafted along in deforming till (Menzies 1989). The exposure of LFA 2 in section OTC 1 displays thrust faulted lower and upper contacts with Cretaceous bedrock and LFA 3 respectively, both contacts plunging below the cliff exposure. This suggests a strong possibility that the segment of LFA 2 is a glacially transported raft of the type described locally by Stalker (1973a, 1976). One fabric on LFA 3 from section OTC 2 displays a NE-SW alignment which is consistent with LFA 3 from elsewhere in the immediate vicinity (Evans and Campbell 1992).

LFA 2 has been displaced over LFA 3

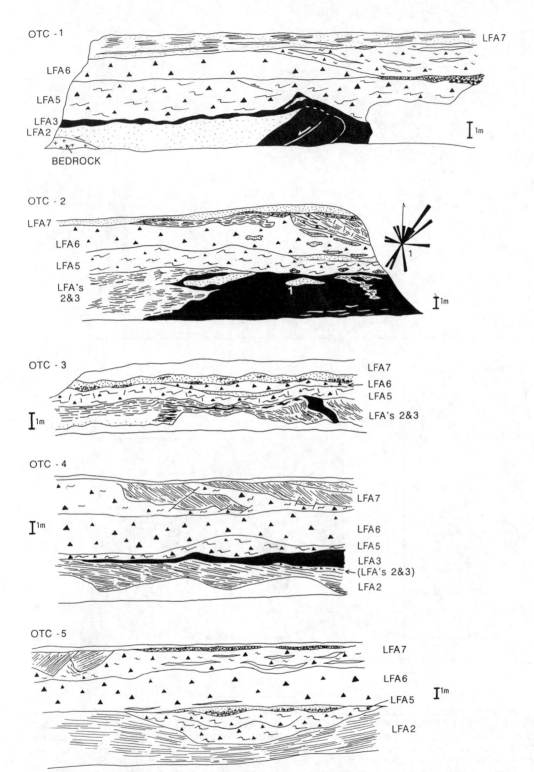

Figure 2: Stratigraphic sections OTC 1-11 exposed along One Tree Creek.
Lithofacies associations (LFA's 1-7) are marked.

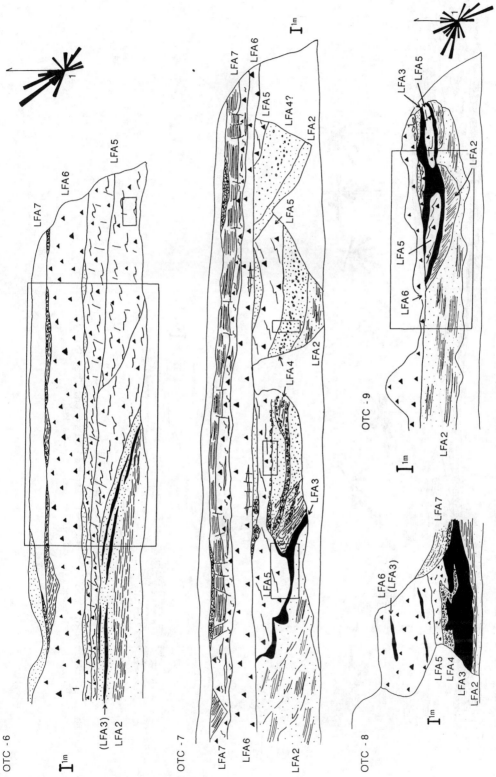

Figure 2: Stratigraphic sections OTC 1–11 exposed along One Tree Creek. Lithofacies associations (LFA's 1–7) are marked.

207

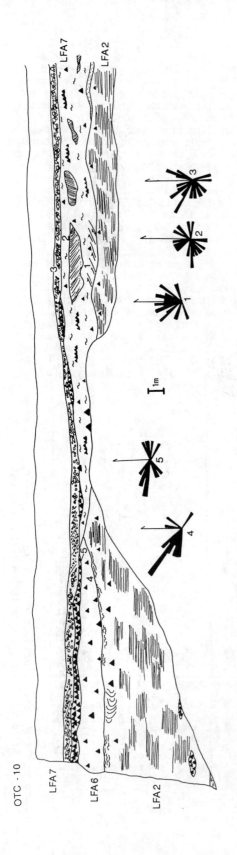

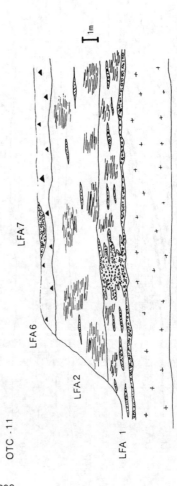

Figure 2: Stratigraphic sections OTC 1-11 exposed along One Tree Creek. Lithofacies associations (LFA's 1-7) are marked.

by thrust faulting in sections OTC 2 and 7 and the interbedding of LFA's 2 and 3 in sections OTC 3 and 6 might represent the attenuation of diamicton where rafts of LFA 2 were imbricately thrust, pinching LFA 3 between thrust plates. The resultant stacked beds were then truncated by glacier erosion. A thin shear zone of intercalated massive diamicton and fines occurs at the base of LFA 3 in section OTC 4 and a sheared melange of silty sand and grey diamicton (LFA's 2 and 3) also occurs in section OTC 6. In some cases the thin interbeds of diamicton and silty sand may be glacitectonic laminations formed at the deforming margins of glacially transported rafts (cf. Hart et al. 1990; Menzies 1989).

To the left of section OTC 2, LFA 2 has been thrust as slices over LFA 3. Rafts of LFA 2 surrounded by isoclinal folds in LFA 3 illustrate primary forms of enclosure of sand and rhythmite bodies by the deforming and shearing till. The axial planes of the isoclinal folds dip generally towards the north. Diapirs in LFA 3 also serve to document intense ductile displacement within the subglacial till caused by its excess loading when unfrozen and its subjection to high porewater pressures. The imbricate thrust faults in LFA's 2 and 4 in section 7, however, indicate that those sediments may have been frozen at the time of glacitectonism. As these facies were being thrust faulted, the till (LFA 3) beneath them was being attenuated and in some places bulldozed by the thrust slices (cf. centre of section 7).

Two general styles of subglacial deformation are identified by Alley et al. (1987) and Hart et al. (1990) based upon evidence from the bases of contemporary and ancient ice sheets (Ice Stream B of Antarctica and the Anglian Ice Sheet of Britain respectively). In areas of compressive flow there is a net thickening of sediment, whereas in areas of extending flow there is a net thinning (cf. Croot 1988). Both the stratigraphic evidence and regional topography dictate compressive flow during ice advance over LFA's 2-4. Their attenuation and shearing combined to create a tectonic melange after initial disruption of LFA 2 by proglacial or ice marginal thrusting along decollement planes. This was most likely akin to the deforming layer of till below Ice Stream B and much like the constructional tectonics envisaged by Hart et al. (1990). Because glacitectonic disturbance affected LFA's 2-4, the deposition of the grey till (LFA 3) most likely dates to a glacial episode which ante-dates the glacitectonic advance. Clearly much of the depositional evidence from the period of LFA 3 and 4 deposition has been removed by the glacitectonic advance, suggesting that glacier bed conditions changed from frozen and compressive to wet based and sliding (extending flow). The erosive contact produced by the sliding ice is best preserved in sections OTC 2-4 and 6. As the One Tree Creek basin presently contains deeply dissected bluffs of preglacial lacustrine sands, silts and clays (LFA 2), it is likely that similar topography confronted the advancing

Figure 3: Aerial view looking east of central One Tree Creek where sections OTC 1-7 provide a valuable three-dimensional exposure of Quaternary stratigraphy.

Laurentide Ice Sheet (Evans and Campbell 1992). Such sediments are relatively permeable compared to underlying Cretaceous strata and overlying LFA 3 (till) and so would be subject to high porewater pressures when groundwater was forced into them by increasing upglacier pressure gradients. This resulted in thrusting at the boundaries of LFA 2 (cf. Bluemle and Clayton 1984). Similar thrust slicing has been observed in proglacial environments and is independent of permafrost conditions (eg. Croot 1988).

4 DEPOSITION OF LFA 5

After the erosive episode which truncated the glacitectonically thrust and deformed LFA's 2-4, there was a period of lodgement as recorded by the deposition of massive diamicton with strong NNE-SSW fabric at the base of LFA 5. This is much the same ice flow direction as that which deposited LFA 3 and glacitectonically deformed LFA's 2-4. The majority of LFA 5 is gravitationally deformed melange of debris flow, fluvial and subaqueous origin which records deposition in subglacial cavities, perhaps during a period of mass stagnation of the ice sheet margin (although the final model does not regard mass stagnation necessary). These sediments display flow features and depositional forms that drape and are controlled by the topography of the glacitectonically disturbed substrate of LFA's 2-4, especially in sections OTC 1, 7 and 8.

Figure 4: Detail of LFA´s 4 and 5 from section OTC 7 showing trough and planar cross-bedded sands and gravels (LFA 4) overlain unconformably by heterogeneous diamicton (LFA 5) containing flow structures, boudins and dropstones.

Figure 5: Details of LFA´s 3-5 from section OTC 7. LFA 3, outlined by broken line, has been attenuated, deformed and overthrust by LFA 4 (stratified sands at right). Both LFA´s 4 and 5 have then been eroded by subglacial meltwater or hyperconcentrated flows and overlain by LFA 5. Sand boudins in LFA 5 (upper left) give impression of mass flow into erosional scours in LFA´s 3 and 4.

Figure 6: Clasts forming a lag in stratified diamicton at the base of LFA 5 in section OTC 7. An erosional contact with the underlying sands of LFA 4 and a lack of any clast long axis alignment indicates erosion of LFA 4 by water or debris flow

Figure 7: Detail of LFA 5 in section OTC 6 showing interbedding of diamicton and sands, flow structures, boudins and normal faulting which are all characteristics of rapid fluvial and mass flow deposition in a subglacial cavity.

Figure 8: Part of section OTC 6 illustrating the largely tabular arrangement of the lithofacies in One Tree Creek and the erosional contact between LFA's 5 and 6. The cavity fill within an erosional scour in LFA(s) 2 (3) is also outlined.

211

Figure 9: Exposure of LFA 7 in the upper reaches of One Tree Creek. Dark muds (outlined) have been loaded by silts/fine sands during the later stages of postglacial lake deposition.

Furthermore, because the lodgement till of lower LFA 5 does not occur in sections OTC 1 and 7 it is considered likely that the heterogeneous diamicton of upper LFA 5 infilled fluvially scoured hollows in the glacitectonically disturbed substrate, thus explaining the removal of the lodgement till (base of LFA 5) beforehand.

5 DEPOSITION OF LFA 6

This diamicton with its strong clast fabric is interpreted by Evans and Campbell (1992) as a till which was deposited by any of a number of processes. Where the diamicton is massive it could have been deposited by melt-out from stationary or active ice, lodgement, or the molding of pre-existing material. Where it is stratified it could have been deposited by melt-out associated with rapid drainage. The depositional reconstruction presented in this paper, based upon the total stratigraphic sequence, appears to favour the molding of pre-existing material followed by melt-out of debris-rich basal ice. At locations outside the One Tree Creek basin, where postglacial spillway floods have not removed all the pertinent evidence (cf. Evans 1991), contorted sand and gravel lenses intercalated with diamicton record final supraglacial deposition of LFA 6 (Evans and Campbell 1992).

Smeared and faulted inclusions of LFA's 2-5 within LFA 6 may have originated from several sources. They are nonetheless critical to the interpretation of the lower part of LFA 6. As is discussed below, it is thought most likely that some inclusions (especially preglacial LFA 2, diamicton LFA 3 and gravel LFA 4) originated as rafts plucked from the substrate by coupled ice and deforming till (Rooney et al. 1987; Menzies 1989). Other lenses of sand and gravel probably originated either as englacial tunnel fills, which were later preserved by melt-out and then deformed by remobilisation of the till, or as conduits within the deforming till layer (pipe flow, Boulton and Hindmarsh 1987). A deformation till origin is preferred for the lower part of LFA 6. Deformation till is defined by Elson (1961, 1989) as: "...derived from beneath the glacier and typically incorporates unconsolidated material such as lake sediments, or weak rocks.... The range of deformation tills encompasses 1) thoroughly homogenised, previously deposited, stratified sediment in which all primary sedimentary structures have been destroyed...; 2) breccia of disoriented fragments retaining some of the original sedimentary structures in a matrix of the same material in a homogenised state; 3) strongly contorted and displaced sediment containing sparse foreign clasts that has moved only a short distance." Alley (1991) has also suggested a deforming-bed origin for the tills of the southern Laurentide Ice Sheet based upon similar characteristics to those outlined by Elson (1961, 1989). A deformation origin explains the sedimentology of lower LFA 6 and is compatible with theoretical ice sheet reconstructions for the western plains which invoke a low profile ice margin coupled with a deforming bed (Boulton and Jones 1979; Fisher et al. 1985).

6 SECOND GENERATION GLACITECTONICS ASSOCIATED WITH LFA's 2-5.

Evidence of a second glacitectonic event occurs in sections OTC 1, 6, 7 and 9. The most convincing evidence of both first and second generation glacitectonics in the One Tree Creek basin is available in sections OTC 7 and 9, both revealing similar sequences to that reported by Campbell and Evans (1990) in the Little Sandhill Creek basin. The second generation glacitectonics occurs to the right of section OTC 7 where a thrust block of LFA's 2, 4? and 5 has been displaced over LFA 5 by ice moving from the north- northeast. In section OTC 9 a recumbent fold in LFA 3 has enclosed pods of LFA 5 and this disturbance appears to have been responsible for internal faulting in underlying LFA 2 (Fig.10). In section OTC 1 a prominent thrust fault developed in LFA 5 is truncated by the erosional contact with LFA 6. This displacement in LFA 5 appears to be related to the protuberance presented by the first generation glacitectonic disturbance in the underlying LFA 3 (grey till). Indeed, the uppermost shear plane observable in LFA 3 could represent an extension of the second generation thrust fault in LFA 5. Numerous normal and thrust faults are associated with an upper sheared zone of LFA 5 in section OTC 6. These indicate disturbance within LFA 5 which was produced by overriding by the coupled ice/till (LFA 6). This overriding produced an erosional contact marked in places by rippled sand lenses which probably record subglacial drainage conduits at the base of the deforming layer (Rooney et al. 1987).

The erosional contact separating LFA's 5 and 6 is interpreted by Evans and Campbell (1992) as the product of renewed contact by this part of the ice sheet margin with its bed after a period of extensive decoupling. The interpretation of LFA 6 (outlined above) is central to the reconstructed glacier dynamics. If the lower part of LFA 6 is a deformation till then it records the flow of glacier ice over a deforming bed. Such a till layer has been detected at the base of Ice Stream B in Antarctica by Alley et al. (1987) and Rooney et al. (1987) who also cite evidence of an erosional contact at the base of the coupled ice/deforming till layer. Evidence of glacitectonic disturbance in the underlying material (in this case LFA's 2-5) has been reported by Rooney et al. (1987) who also suggest that rafts may be entrained in the deforming layer. This would explain the preservation of large faulted blocks of preglacial sediment in LFA 6 (Evans and Campbell 1992; Menzies 1989) and would provide an alternative mechanism of mega-block entrainment to those proposed by Moran et al. (1980) and Bluemle and Clayton (1984).

It is proposed that the second generation glacitectonics in LFA's 2-5 in sections OTC 1, 6, 7 and 9 were produced by the shear stress applied by the coupled ice/till (lower LFA 6) when the glacier renewed contact with its bed after a period of subglacial meltwater production and cavity sedimentation. Therefore, the deforming till layer would have been composed of remobilised LFA 5 which, prior to its erosion by the coupled ice/till, possessed an irregular surface due to its original deposition into subglacial cavities. Where the upper beds of LFA 6 are present (for example in the Little Sandhill Creek area), Evans and Campbell (1992) have used shear planes and clast fabrics to suggest later remobilisation of the diamicton by ice flow from the northeast or east as opposed to the general NW-SE fabric of lower LFA 6. Evans and Campbell go on to suggest that ice further up the flow line was surging and bulldozing the stagnant ice located over the field area.

7 DEPOSITION AND GRAVITY TECTONICS OF LFA 7a

It is important to include LFA 7a in this paper because it displays examples of tectonics and rafts included in diamictons which could be misinterpreted as pro-/subglacial in origin. Particularly good examples occur in sections OTC 1, 2, 4-6, 7 and 10. In sections 3, 8 and 9, LFA 7 is dominated by postglacial alluvium and much of LFA 7a has been removed. Evans and Campbell (1992) interpreted LFA 7a as proglacial lacustrine sediment. Where blocks of rhythmites are suspended in or are in fault contact with stratified diamicton they are considered indicative of subaqueous mass failure in the

Figure 10: Overfold in section OTC 9 where LFA 3 has been wrapped around LFA 5 providing evidence of a post-LFA 5 glacitectonic event. Geological hammer circled for scale.

proglacial lake environment (cf. Eyles 1987). Clast fabrics from LFA's 6 and 7a in section OTC 10, where LFA 7a is superimposed on and replaces LFA 6, suggest a subaqueous rather than a subglacial depositional environment. Also in section OTC 10, small shear planes at the base of LFA 7a and internal gravel lags confirm a subaqueous mass flow origin. In sections OTC 1, 2, 5 and 7 clay beds conforming to the surface topography of LFA 6 provide further evidence that LFA 7a was deposited in a lacustrine environment. Furthermore, intense disturbance indicative of water escape and sediment loading occur in an upstream exposure of LFA 7a (Fig. 9).

8 GLACIAL PROCESS-FORM AND LANDSYSTEMS MODELS

Such models have been proposed for the glaciated prairie landscape and they assume that former glacial processes and their spatial and temporal variations can be deduced largely from landform assemblages and patterns (Clayton and Moran 1974; Mooers 1990). Landsystems models, on the other hand, use sub-surface glacial geology to determine changing styles of glacigenic deposition through time (Eyles 1983).

In the case of One Tree Creek, the generally flat prairie surface, which is product of the underlying proglacial lake

sediments, completely masks the complexity of glacial erosional, depositional and tectonic process-form evidence. Clayton and Moran (1974) emphasised a need to explain glacial landforms in terms of several process continuums and to do this the glacial geologist must be familiar with the stratigraphy of sites under study. They also emphasised the importance of glacier bed temperatures and groundwater pressures in the substrate. Basal thermal regimes have been considered in several process-form models and have been regarded by many authors as critical in the explanation of glacial tectonics (cf. Boulton 1972; Aber 1982, 1985; Bluemle and Clayton 1984; Mooers 1990).

It is with all these provisos in mind that a glacial process model has been constructed for the One Tree Creek basin based upon the stratigraphic information outlined above. This model clearly has major implications for the reconstruction of former ice dynamics at the southwest margin of the Laurentide Ice Sheet. A complete landsystems approach, combining the stratigraphic data from the One Tree Creek/Dinosaur Provincial Park area (cf. Campbell and Evans 1990; Evans and Campbell 1992) with the wider scale glacial geomorphology, will be presented in a later paper.

214

9 A MODEL OF FORMER ICE DYNAMICS OVER THE ONE TREE CREEK BASIN

During the last (Lostwood) glaciation, glacier ice advanced over the One Tree Creek area and, due to the compressive flow enforced by the high land to the south, thrust coarse-grained proglacial sediments (LFA 2), grey till (LFA 3) and sand and gravel outwash (LFA 4). Glacitectonic thrusting was most likely initiated proglacially and then continued in the marginal subglacial zone (Fig.11,i).

The grey till (LFA 3) responded to glacier stress very differently to LFA's 2 and 4 by undergoing ductile folding and attenuation. There are two possible explanations for this: First, if the proglacial zone was permafrozen, coarse-grained sediments would have retained much of their integrity even though they became thrust and stacked. Because LFA 3 is clay rich, it would have remained unfrozen even at sub-zero temperatures and therefore would have been subject to deformation (Mathews and Mackay 1960; Mackay and Mathews 1964). Second, if the advancing glacier ice caused the damming of considerable amounts of lake water then permafrost development would have been prohibited beneath deep water. Such a scenario is envisaged by Tsui et al (1989) who have noted an association of many ice thrust terrains in Alberta with preglacial valleys. The inclination of regional slope towards the advancing ice margin would compound this process. Therefore, thrust faulting would have occurred in the less competent beds of LFA's 2 and 4 and deformation of LFA 3 resulted from excess shear stress being applied to more competent, clay rich sediment (Croot 1988).

Regardless of which explanation is correct, the overriding of the site by glacier ice caused the deformation of till and its folding around older (formerly underlying) silt/clay thrust slices as they were displaced vertically into overlying till and were also horizontally shortened. A similar example of the buckling of upper deformable beds by displaced lower beds occurs on the Norfolk coast of England at Trimingham where chalk blocks have been thrust upwards into stratified sediments and diamicton (cf. Eyles et al 1989, Fig.3).

After the advance of the snout zone past One Tree Creek, the glacitectonic disturbance phase was followed by a period of subglacial deformation when the lodgement component of LFA 5 was deposited (Fig.11, ii). It is possible that this material is in fact a deformation till, indicating a coupled ice/till layer but exposures are restricted and it is not possible to reject a lodgement (sliding bed) origin. However, the change from proglacial thrusting and stacking to basal sliding or ice/till deformation records a change in basal thermal regime related to the advance of the cold based snout well beyond the site.

Following the lodgement or deformation phase there was a period of subglacial cavity excavation and filling by meltwater, explaining the deposition of LFA 5 (Fig.11, iii). This depositional phase was related to the channelling or piping of subglacial meltwater and marked a period of increased meltwater discharge. Boulton and Hindmarsh (1987) have modelled the initiation of subglacial drainage patterns as a response to increasing meltwater discharge and have concluded that substrate deformation would continue at a stable rate. Instability would accrue if the outflow of subglacial meltwater from the glacier system was impeded by a frozen snout zone, as would appear to be the case in southern Alberta. This would result in surging behaviour (Clarke et al. 1984).

Such instability and a surge(s) would explain the structures displayed by sediments (LFA's 1-5) in One Tree Creek (Fig.11, iv). Although only one major phase of such instability appears to be recorded in One Tree Creek, further periods of subglacial meltwater discharge and consequent surge activity were probably centred over other adjacent areas. That such activity continued is evident in the structures of LFA 6 (Fig.11, iv-vii) which record a period of ice/till deformation (Fig.11, iv) followed by ice stagnation and reactivation (Fig.11, v-vii).

The subdivision of LFA 6 is based upon sediments and structures produced by changes in the activity and basal thermal regime of the glacier ice over One Tree Creek. The preservation of attenuated rafts of LFA's 1-5 in the lower massive, sheared diamicton records ice/substrate coupling (Fig.11, iv); the diamicton is therefore a deformation till. A subsequent period of ice stagnation produced a partial melt-out sequence of interbedded gravels, sands and diamictons (Shaw 1982, 1987; Fig.11, v). This was followed by at least one ice reactivation phase which produced a massive, sheared diamicton with attenuated gravel and sand lenses (another deformation till; Fig.11, vi). That this reactivation and remobilisation of glacigenic sediments continued as a result of up-ice surging is verified by till fabrics on LFA 6 outside the One Tree Creek area (Campbell and Evans 1990; Evans and Campbell 1992).

The final phase of deposition for LFA 6 was that of melt out and supraglacial flowage, producing contorted, lensoid bodies of sand and gravel suspended in diamictons (Fig.11, vii and viii). Much of this component of LFA 6 has been removed in One Tree Creek but the subsequent proglacial lacustrine deposits are extensive (Fig.11, viii).

10 DISCUSSION

Work on contemporary ice sheets has revealed a close association between ice streaming and deforming substrates (Alley et al. 1986, 1987; Rooney et al. 1987). Furthermore, ice formerly advancing from the Keewatin and Hudson Bay sectors of the Laurentide Ice Sheet has taken abrupt directional shifts after moving from rigid (Shield) beds with sporadic surficial cover to deformable Cretaceous and Tertiary beds with a considerable cover of unlithified Quaternary sediments (Dyke et al 1982; Fisher et al 1985). Evidence of such ice sheet behaviour is available in the glacial geomorphology of Alberta (cf.

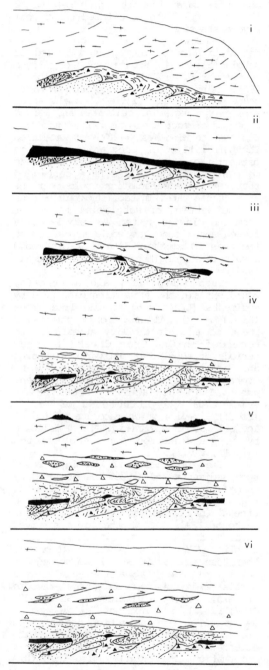

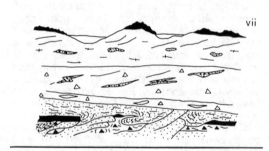

Figure 11 Model of glacier dynamics and associated sediments and structures over One Tree Creek. This sequence appears to satisfactorily explain the sediments and structures found in the One Tree Creek sections.

GLACIER ICE WITH FOLIA

LFA's 1 AND 2

LFA 3

LFA 4

LOWER LFA 5

UPPER LFA 5

SUBGLACIAL WATER FLOW

LOWER LFA 6
[DEFORMATION TILL]

MIDDLE LFA 6 [PRECURSOR]
MELT-OUT TILL

MIDDLE LFA 6 WITH SHEAR PLANES
[DEFORMATION TILL]

STAGNANT ICE WITH SUPRAGLACIAL
DEBRIS AND PONDS AND ENGLACIAL
STREAM GRAVELS

CONTORTED GRAVEL AND SAND
LENSES OF UPPER LFA 6

LFA 7

Figure 11: Model of glacier dynamics and associated sediments and structures over One Tree Creek (cont.).

Evans 1985; Shetsen 1987, 1990; Klassen 1989) and southern Ontario (Boyce and Eyles 1991) where ice directional indicators such as drumlins and flutings reveal areas of ice streaming (Fig.12). Classic examples of such streaming are the "Innisfail Lobe" of Stalker (1973b) and the Lac la Biche fluting field of Jones (1982). In the latter case the geomorphic and stratigraphic evidence indicates that an initial, cold based glacier advance phase was responsible for glacitectonic thrusting and this was followed by a warm based ice streaming phase.

A corollary is that if thrust terrains are formed during ice advance then their survival as landforms attests to areas of inactive ice during later overriding. This may be tested by an analysis of the distribution of thrust terrains and palaeo-ice streams. That many of the ice thrust terrains of Alberta do date to initial ice advance is indicated by the fact that they have been cross-cut by subglacial meltwater channels which are in turn draped by hummocky moraine (Klassen 1989). It is distinctly possible that such channels were cut during a subglacial cavity fill phase similar to that which occurred in One Tree Creek. The ubiquity of subglacial meltwater under the southwestern part of the Laurentide Ice Sheet is illustrated by the, apparently fluvially derived, erosional and depositional forms of Shaw (1989) and Shaw et al (1989). Shifting reservoirs of subglacial meltwater may have had profound effects on the distribution of both fluvially and glacially derived flutings.

Although large volumes of subglacial meltwater appear to have been available in the western plains, it is unlikely that the processes of tunnel valley cutting, cavity filling, ice streaming/till deformation and glacier surging were strictly contemporaneous. In fact it is more likely that the initiation of one process in a specific area would set off a chain reaction elsewhere. Evidence for the time-transgressive progression of stages i-vii (Fig.11) in the One Tree Creek model is manifest not only in the stratigraphic data but also in the geomorphology of the nearby plains. For example, former subglacial tunnel valleys which occur on the prairie surface north of Dinosaur Provincial Park (cf. Evans and Campbell 1992) provide evidence of continued meltwater activity to the northeast of One Tree Creek during stages vi-viii (Fig.11), suggesting that subglacial piping had shifted eastwards after stage iii (Fig.11).

11 CONCLUSIONS

It is apparent from the study of stratigraphic sections in the One Tree Creek basin that sub-surface sedimentary sequences on the western plains of Canada can be extremely complex and as a result depositional reconstructions of multiple "tills" should be re-assessed. However, this study does demonstrate that a lithofacies approach to Quaternary stratigraphy in Alberta can provide valuable information on depositional reconstructions (cf. Shaw 1987) and, together with information provided by

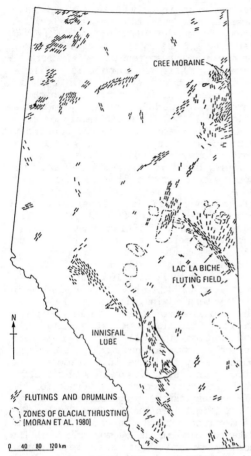

Figure 12 Map of Drumlins and flutings in Alberta depicting areas of fast glacier flow.

recent studies on contemporary glaciers (cf. Alley et al 1986, 1987; Boulton and Hindmarsh 1987; Rooney et al 1987), has much to contribute to the modelling of palaeo-ice sheets (cf. Fisher et al 1985). It is also possible that some thick multiple-stacked sequences of diamictons, previously interpreted as lodgement tills, melt-out tills or supraglacial diamicts (Shaw 1982; Paul and Eyles 1990; Evans and Campbell 1992), are in fact extensive deformation tills and as such are a valuable source of information for the reconstruction of the poorly understood southwestern margin of the Laurentide Ice Sheet.

ACKNOWLEDGEMENTS

This research was funded by a University of London, Central Research Fund grant. Indirect funding came from a Natural Sciences and Engineering Research Council of Canada grant awarded to Dr I.A. Campbell, University of Alberta while the author was undertaking postdoctoral research. Dr Campbell is also acknowledged for his continuing support and collaboration. Thanks to M. Seemann and D. Chesterman for field support and to the local landowners and the Eastern Irrigation District for kind permission to access One Tree Creek.

REFERENCES

Aber J.S. (1982) Model for glaciotectonism. Geological Society of Denmark Bulletin, 30: 79-90.
Aber J.S. (1985) The character of glaciotectonism. Geologie en Mijnbouw, 64: 389-395.
Alley R.B. (1991) Deforming-bed origin for southern Laurentide till sheets? Journal of Glaciology, 37: 67-76.
Alley R.B., Blankenship D.D., Bentley C.R. and Rooney S.T. (1986) Deformation of till beneath Ice Stream B, West Antarctica. Nature, 322: 57-59.
Alley R.B., Blankenship D.D., Bentley C.R. and Rooney S.T. (1987) Till beneath Ice Stream B, 3. till deformation: evidence and implications. Journal of Geophysical Research, 92: 8921-8929.
Bluemle J.P. and Clayton L. (1984) Large scale glacial thrusting and related processes in North Dakota. Boreas, 13: 279-299.
Boulton G.S. (1972) The role of the thermal regime in glacial sedimentation. Institute of British Geographers, Special Publication 4: 1-20.
Boulton G.S. and Hindmarsh R.C.A. (1987) Sediment deformation beneath glaciers: rheology and geological consequences. Journal of Geophysical Research, 92: 9059-9082.
Boulton G.S. and Jones A.S. (1979) Stability of temperate ice caps and ice sheets resting on beds of deformable sediment. Journal of Glaciology, 90: 29-43.
Boyce J.I. and Eyles N. (1991) Drumlins carved by deforming till streams below the Laurentide ice sheet. Geology, 19: 787-790.
Campbell I.A. and Evans D.J.A. (1990) Glaciotectonism and landsliding in Little Sandhill Creek, Alberta. Geomorphology, 4: 19-36.
Clarke G.K.C., Collins S.G. and Thompson D.E. (1984) Flow, thermal structure, and

subglacial conditions of a surge-type glacier. Canadian Journal of Earth Sciences, 21: 232-240.

Clayton L. and Moran S.R. (1974) A glacial process form model. In, Coates D.R. (ed), Glacial Geomorphology: State University of New York, Binghamton, 89-119.

Clayton L., Teller J. and Attig J. (1985) Surging of the southwestern part of the Laurentide Ice Sheet. Boreas, 14: 235-241.

Croot D.G. (1988) Morphological, structural and mechanical analysis of neoglacial ice-pushed ridges in Iceland. In, Croot D.G. (ed.), Glaciotectonics: Forms and Processes. Balkema, Rotterdam, p.33-47.

Dyke A.S., Dredge L.A. and Vincent J-S. (1982) Configuration of the Laurentide ice sheet during the Late Wisconsin maximum. Geographie Physique et Quaternaire, 36: 5-14.

Elson J.A. (1961) The geology of tills. Proceedings of the 14th Soil Mechanics Conference. National Research Council of Canada, Associate Committee on soil and snow mechanics, Technical Memorandum No.69, Ottawa, p.5-35.

Elson J.A. (1989) Comment of glacitectonite, deformation till, and comminution till. In, Goldthwaite R.P. and Matsch C.L. eds., Genetic Classification of Glacigenic Deposits. Balkema, Rotterdam, p.85-88.

Evans D.J.A. (1985) Wisconsin ice dynamics in Alberta: A review of prevalent hypotheses and the development of recent ideas. In, Deshaies L. and Pelletier R. (eds), Proceedings of the Canadian Association of Geographers, Trois Rivieres, 13-33.

Evans D.J.A. (1991) A gravel/diamicton lag on the south Albertan prairies, Canada: evidence of bed armouring in early deglacial sheet-flood/spillway courses. Geological Society of America Bulletin, 103: 975-982.

Evans D.J.A. and Campbell I.A. (1992) Glacial and postglacial stratigraphy of Dinosaur Provincial Park and surrounding plains, southern Alberta, Canada. Quaternary Science Reviews.

Eyles N. (1983) Glacial geology: a landsystems approach. In, Eyles N. (ed), Glacial Geology: Pergamon, 1-18.

Eyles N. (1987) Late Pleistocene debris flow deposits in large glacial lakes in British Columbia and Alaska. Sedimentary Geology, 53: 33-71.

Eyles N., Eyles C.H. and McCabe A.M. (1989) Sedimentation in an ice-contact subaqueous setting: the mid-Pleistocene "north sea drifts" of Norfolk, U.K.. Quaternary Science Reviews, 8: 57-74.

Fisher D.A., Reeh N. and Langley K.(1985) Objective reconstructions of the Late Wisconsinan ice sheet and the significance of deformable beds. Geographie Physique et Quaternaire, 39: 229-238.

Hart J.K., Hindmarsh R.C.A. and Boulton G.S. (1990) Styles of subglacial glaciotectonic deformation within the context of the Anglian ice sheet. Earth Surface Processes and Landforms, 15: 227-241.

Jones N. (1982) The formation of glacial flutings in east-central Alberta. In, Davidson-Arnott R., Nickling W. and Fahey B.D. (eds), Research in Glacial, Glaciofluvial and Glaciolacustrine Systems. Geobooks, Norwich, 49-70.

Klassen R.W. (1989) Quaternary geology of the southern Canadian interior plains. In, Fulton R.J. (ed), Quarternary Geology of Canada and Greenland. Geological Society of America, The Geology of North America, Vol.K-1, 138-173.

Kupsch W.O. (1962) Ice-thrust ridges in western Canada. Journal of Geology, 70: 582-594.

Mackay J.R. and Mathews W.W. (1964) The role of permafrost in ice thrusting. Journal of Geology, 72: 378-380.

Mathews W.W. and Mackay J.R. (1960) Deformation of soils by glacier push and the influence of pore pressures and permafrost. Transactions of the Royal Society of Canada, 54: 27-36.

Menzies J. (1989) Subglacial hydraulic conditions and their possible impact upon subglacial bed formation. Sedimentary Geology, 62: 125-150.

Mooers H.D. (1990) A glacial-process model: the role of spatial and temporal variations in glacier thermal regime. Geological Society of America Bulletin, 102: 243-251.

Moran S.R. (1971) Glaciotectonic structures in drift. In, Goldthwait R.P. (ed), Till: A Symposium: Ohio State University Press, Columbus, 127-148.

Moran S.R., Clayton L., Hooke R. LeB., Fenton M.M. and Andraishek L.D. (1980) Glacier bed landforms of the prairie region of North America. Journal of Glaciology, 25: 457-476.

Paul M.A. and Eyles N. (1990) Constraints on the preservation of diamict facies (melt-out tills) at the margins of stagnant glaciers. Quaternary Science Reviews, 9: 51-69.

Rooney S.T., Blankenship D.D., Alley R.B. and Bentley C.R. (1987) Till beneath Ice Stream B, 2. structure and continuity. Journal of Geophysical Research, 92: 8913-8920.

Shaw J. (1982) Melt out till in the Edmonton area, Canada. Canadian Journal of Earth Sciences, 19: 1548-1569.

Shaw J. (1987) Glacial sedimentary processes and environmental reconstruction based on lithofacies. Sedimentology, 34: 103-116.

Shaw J. (1989) Drumlins, subglacial meltwater floods, and ocean responses. Geology, 17: 853-856.

Shaw J., Kvill D. and Rains R.E. (1989) Drumlins and catastrophic subglacial floods. Sedimentary Geology, 62: 177-202.

Shetsen I. (1987) Quaternary Geology, Southern Alberta. 1:500,000 scale map, Alberta Research Council.

Shetsen I. (1989) Quaternary Geology, Central Alberta. 1:500,000 scale map, Alberta Research Council.

Slater G. (1927) Structure of the Mud Buttes and Tit Hills in Alberta. Geological Society of America Bulletin, 38: 721-730.

Stalker A. MacS. (1973a) The large interdrift bedrock blocks of the Canadian prairies. Geological Survey of Canada, Paper 75-1A, 421-422.

Stalker A. MacS. (1973b) Surficial Geology of the Drumheller Area. Geological Survey of Canada, Memoir 370.

Stalker A. MacS. (1976) Megablocks, or the enormous erratics of the Albertan prairies. Geological Survey of Canada, Paper 76-1C, 185-188.

Tsui P.C., Cruden D.M. and Thomson S. (1989) Ice-thrust terrains and glaciotectonic settings in central Alberta. Canadian Journal of Earth Sciences, 26: 1308-1318.

Formation and Deformation of Glacial Deposits, Warren & Croot (eds) © 1994 Balkema, Rotterdam, ISBN 90 5410 096 6

List of contributors

Pete Coxon
Department of Geography,
Trinity College,
Dublin 2,
Republic of Ireland

George F. Dardis
Sedimentology and Palaeobiology Laboratory,
Anglia Polytechnic,
East Road,
Cambridge, CB1 1PT

Dietrich Ellwanger
Geologishces Landesamt Baden-Württemberg,
Albertstr. 5
D-79104 Freiburg i.Br.

David J.A. Evans
Dept of Geography and Topographic Science
University of Glasgow
Glasgow G12 8QQ
Scotland

Joanna M.R. Fernland
Royal Institute of Technology
Engineering Geology
S-100 44
Stockholm

Patricia M. Hanvey
Department of Geography and Env. Studies,
University of the Witwatersrand,
1 Jan Smuts Ave, Johannesburg
South Africa

Peter Johnansson
Geological Survey of Finland

Dariusz Krzyszkowski
Geographical Institute
University of Wroclaw
Poland

Eric Lagerlund
Dept of Quaternary Geology,
Lund University
Lund
Sweden

Jaap J.M. van der Meer
Fysisch Geografisch en Bodemkundig Laboratorium
University of Amsterdam
Nieuwe Prinsengracht 130
1018 VZ Amsterdam

Lewis A. Owen
Royal Holloway and Bedford New College

Carrie J. Patterson
Minnesota Geological Survey, St Paul
Minnesota, USA

Kärstin Malmberg Persson
Dept of Quaternary Geology
University of Lund, Sweden

Jan A. Piotrowski
Institute of Geology and Palaeontology
University of Kiel
Kiel, Germany

Thorsteinn Saemundsson
Dept of Quaternary Geology
University of Lund, Sweden

J Saettem
IKU Petroleum Research
Norway

Lennart Sorby
Dept of Quaternary Geology
University of Lund
Sweden

R.R. Stea
Nova Scotia Dept of Natural Resources
PO Box 698
Halifax, Nova Scotia
Canada BBJ 2T9

Anja L.L.M. Verbers
Fysisch Geografisch en Bodemkundig
Laboratorium
University of Amsterdam
Nieuwe Prinsengracht 130
1018 VZ Amsterdam

William P. Warren
Geological Survey of Ireland,
Beggars Bush, Haddington Rd
Dublin 4, Ireland

Formation and Deformation of Glacial Deposits, Warren & Croot (eds) © 1994 Balkema, Rotterdam, ISBN 90 5410 096 6

Author index